Melt Pool In-Situ Controlling of Laser Additive Manufacturing Technology

Kaiyu Luo • Youyu Su • Jinzhong Lu

Melt Pool In-Situ Controlling of Laser Additive Manufacturing Technology

Springer

Kaiyu Luo
School of Mechanical Engineering
Jiangsu University
Zhenjiang, Jiangsu, China

Youyu Su
School of Mechanical Engineering
Nantong University
Nantong, Jiangsu, China

Jinzhong Lu
School of Mechanical Engineering
Jiangsu University
Zhenjiang, Jiangsu, China

ISBN 978-981-95-4596-4 ISBN 978-981-95-4597-1 (eBook)
https://doi.org/10.1007/978-981-95-4597-1

This Springer imprint is published by the registered company Springer Nature Singapore Pte Ltd.
The registered company address is: 152 Beach Road, #21-01/04 Gateway East, Singapore 189721, Singapore

Preface

Since its emergence in the 1970s, laser metal additive manufacturing technology has achieved significant progress in fundamental theoretical research. However, performance instabilities during continuous processing—such as thermal history effects, metallurgical defects, and epitaxial growth of columnar grains—have consistently hindered its widespread adoption in the standardized industrial production.

The melt pool in additive manufacturing processes features high temperature, high brightness, small size, and rapid changes, generated by high-energy laser irradiation on the substrate. The sequential arrangement and accumulation of melt pools induce thermal accumulation reactions, necessitating real-time monitoring and regulation. Existing research indicates that online monitoring and in-situ control of the melt pool can effectively enhance component forming quality and mechanical properties.

This book details the application of in-situ melt pool control-assisted laser directed energy deposition in high-performance manufacturing of critical components. However, challenges persist in the synergistic control of forming quality and performance, primarily involving dimensional accuracy and microstructure regulation issues caused by cumulative thermal effects. This book systematically compares traditional LDED and LDED with in-situ melt pool control in terms of surface quality, microstructure, and overall performance for single-track, single-layer, and thin-walled 316L components. It highlights core research findings, laying the theoretical foundation for elucidating the principles and effects of thermo-mechanical interactions in laser metal additive manufacturing. The book is divided into four parts, comprehensively presenting laser metal additive manufacturing technology based on thermo-mechanical interactions. The first part reviews the background of melt pool in-situ control technology in laser metal additive manufacturing. The second part elaborates on the theoretical foundations of LDED and closed-loop control technology. The remaining two parts specifically describe the microstructural evolution, forming quality, and overall performance of materials when employing in-situ melt pool control technology.

This book holds significant value for readers in mechanical engineering and materials science, as in-situ melt pool control-assisted laser metal additive manufacturing is particularly suited for manufacturing and applying critical aerospace engine components. The analyses herein will deepen readers' understanding of the technology's mechanisms. The author's unique insights into thermomechanically coupled laser metal additive manufacturing permeate the text, offering potential guidance for the technology's future advancement.

The research work from which this book arises was carried out at School of Mechanical Engineering in Jiangsu University supported by the National Natural Science Foundation of China (Grant Numbers: U2570239 and 52305469), Jiangsu Provincial Science and Technology Projects in China (Grant Numbers: BE2023026 and BE2022069). Professor Kaiyu Luo at Jiangsu University finished Chaps. 1 and 5–8 , Dr. Youyu Su at Nantong University finished Chaps. 2–4 and 8, and Professor Jinzhong Lu at Jiangsu University finished Chap. 1. The authors would like to thank their colleagues and friends for useful discussions and help in the preparation of this book.

Due to our limited knowledge and energy, there inevitably exist some ambiguous interpretations and even mistakes in this book, which we welcome the readers and colleagues to point out.

Zhenjiang, Jiangsu, China — Kaiyu Luo
Nantong, Jiangsu, China — Youyu Su
Zhenjiang, Jiangsu, China — Jinzhong Lu

Contents

Chapter 1
General Introduction

1.1 Introduction Laser Directed Energy Deposition

1.1.1 Background of Additive Manufacturing

The aerospace, petroleum and mining, nuclear power, and transportation industries, as strategic high-tech industries in China's national defense strategy, are important indicators of a nation's scientific and technological development and economic growth. They also play a crucial role in advancing national defense modernization and enhancing a country's overall strength. Among these, the development of the aerospace industry holds significant importance for economic growth, national security, scientific and technological innovation, and transportation, serving as a key indicator of a nation's comprehensive strength and competitiveness. The aircraft engine, often referred to as the "heart of aviation equipment," is hailed as the "flower of modern industry". Improving the overall performance of key components of aviation engines and reducing their failure rates are critical to extending the service life of aerospace equipment [1–3].

316L stainless steel has a low thermal expansion coefficient and excellent corrosion resistance, and is currently primarily used in the aerospace industry for components such as engine combustion chambers, turbine discs, blades, and complex internal engine parts [4, 5]. Using 316L stainless steel ensures a longer service life and higher reliability for aviation engines. Traditional manufacturing methods for 316L stainless steel aviation engine critical components include molding, casting, and machining processes, which not only involve complex structural designs and high costs but also require each detail to undergo traditional tooling, fixtures, and multiple machining processes [6, 7], resulting in time-consuming, labor-intensive operations that increase production time and costs.

With the upgrading of modern manufacturing processes, the heavy-duty equipment is rapidly evolving toward larger-scale, integrated, and high-performance

K. Luo et al., *Melt Pool In-Situ Controlling of Laser Additive Manufacturing Technology*, https://doi.org/10.1007/978-981-95-4597-1_1

designs. This trend has placed increasingly stringent demands on the high-quality and efficient manufacturing technologies required for large-scale metal structural components. As shown in Fig. 1.1, critical components such as large-scale metal stamping dies, aerospace aircraft turbine blades, and hydraulic support columns for coal mines are subject to harsh operating conditions, making them highly susceptible to environmental corrosion and damage, which can lead to component failure. This not only wastes materials but also increases production costs, necessitating the development of a new processing method.

Under these production demands, rapid prototyping technology has begun to develop rapidly. It is a new processing technology that integrates laser technology, computer technology, automation technology, image processing technology, optical detection technology, and chemical materials, featuring short processing cycles, low material costs, diverse forming structures, and rich processing modes [8, 9].

Laser additive manufacturing technology emerged as a new processing technique in the mid-1980s. The first patent for 3D printing was created in 1984 by Chuck Hull of 3D Systems Corporation [10]. Hull defined the 3D printing process as a system that generates 3D objects by creating cross-sectional patterns of the object to be formed. Metal 3D printing emerged in the 1990s with the invention of laser melting and sintering technologies. In 2010, the ASTM F42 Additive Manufacturing Task Force of the American Society for Testing and Materials (ASTM) proposed a set of standards, classifying additive manufacturing processes into seven categories. ISO/ASTM 52900-15 defines seven categories of additive manufacturing processes: Material Extrusion, Photopolymerization, Powder Bed Fusion, Material Jetting, Binder Jetting, Sheet Lamination, and Directed Energy Deposition.

Laser additive manufacturing employs a point-by-point and layer-by-layer scanning method, achieving the integrated formation of three-dimensional solids through layered stacking. Within the melt pool, there is heat exchange and morphological changes [11, 12]. Due to its ability to rapidly and precisely manufacture complex-shaped structural components, low manufacturing costs, and high forming efficiency, laser additive manufacturing technology demonstrates significant advantages in the rapid forming of large-sized and complex parts. As a result, it holds broad application prospects in fields such as aerospace, automotive, and maritime industries. Laser additive manufacturing processes using powder as raw material can be categorized based on the interaction between powder and laser during the additive

Fig. 1.1 Physical drawings of large equipment such as large metal stamping die, turbine blade, and coal mine hydraulic support

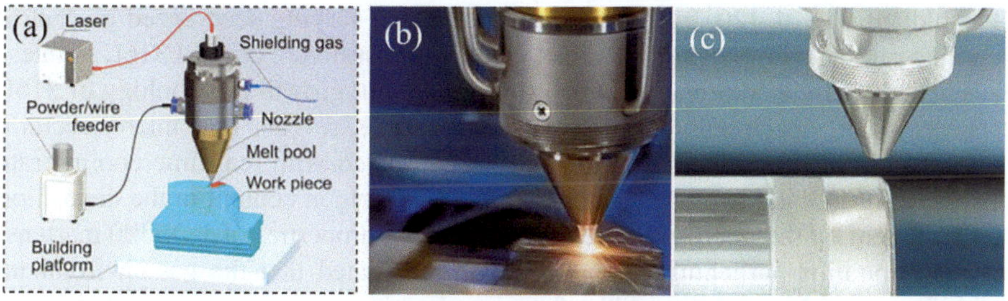

Fig. 1.2 (**a**) The forming principle diagram of laser directed energy deposition technology; (**b**) laser directed energy deposited flat parts; (**c**) laser directed energy deposited tube

process and the powder feeding method. These include Laser Powder Bed Fusion (LPBF) [13], which employs a powder spreading method, and Laser Directed Energy Deposition (LDED) [14], which uses a coaxial powder feeding method.

1.1.2 Principle and Characteristics of LDED

Laser directed energy deposition (LDED) differs from LPBF in that it features simultaneous powder feeding. As shown in Fig. 1.2, the laser head of an LDED device directs a high-energy laser beam onto the substrate, melting the powder to form a melt pool, thereby achieving metallurgical bonding with the substrate. The LDED process can be used to deposit various materials such as stainless steel, titanium alloys, and cobalt-chromium alloys [15, 16]. As a novel surface treatment technology, LDED has a minimal thermal impact on the substrate and high material utilization, making it widely applied in manufacturing fields [17–19]. Additionally, compared to LPBF technology, LDED utilizes higher laser power and larger laser beam sizes to achieve higher deposition efficiency.

1.2 Current Development Status and Challenges

The forming quality and overall performance of LDED-manufactured parts generally depend on the following factors: (1) the type of deposition technology; (2) the build environment (vacuum, inert gas, or ambient); (3) interactions between deposition materials; (4) deposition parameters (powder particle size, scanning speed, powder feed rate, scanning path, overlap ratio, etc.); (5) the type of powder feedstock (titanium alloys, stainless steels, nickel-based superalloys, magnesium-aluminum alloys, etc.). Furthermore, during the layer-by-layer deposition process, LDED-deposited parts undergo rapid, repetitive heating-cooling cycles, resulting in unique microstructural features, non-equilibrium phases, solidification cracking, directional solidification, residual stresses, porosity, delamination, and warpage.

Farshidianfar et al. [20, 21] proposed a method of varying scan speed to control cooling rates and investigated its effect on the microstructure of AISI 316L stainless steel produced via additive manufacturing. The grain growth morphology of 316L stainless steel specimens at different cooling rates reveals that microstructural changes during solidification are positively correlated with real-time cooling rate variations. Li et al. [22] investigated the effect of oxygen content in the deposition environment on the microstructure and mechanical properties of 18Ni300 martensitic precipitation-hardening steel. The results indicated that the gas environment significantly influenced the width of equiaxed grain regions, grain growth orientation, and primary dendrite arm spacing. In the low-oxygen group, higher microhardness values were obtained due to a heat treatment-like effect. Both 500 ppm and 50 ppm oxygen concentrations effectively improved specimen forming accuracy and process stability. Xu et al. [23] compared the microstructure and corrosion resistance of Cr-rich stainless steel coatings produced via conventional-rate deposition and ultra-high-speed laser deposition, finding that increased scanning speed reduced porosity and enhanced corrosion resistance. Su et al. [24] fabricated TC4/316L stainless steel functional gradient materials with varying premix ratios, characterizing and investigating the effects of different TC4/316L ratios on the alloy's microhardness, tensile properties, and wear performance.

Constrained by the aforementioned forming conditions, LDED faces the following challenges in practical processing: (1) Local temperature differences may cause shrinkage, residual stresses, and deformation; (2) Compared to LPBF, it exhibits lower dimensional resolution (precision) and greater surface waviness; (3) Within the powder feeding system, surface roughness is higher than that of LPBF; (4) Part complexity may be constrained, particularly in machines limited to three degrees of freedom; (5) Post-processing is typically required; (6) Powder efficiency and powder recyclability are lower than LPBF, especially when printing powder mixtures. Generally, due to the directionality of deposition, LDED specimens often exhibit anisotropy in mechanical properties and non-uniform microstructure. Consequently, the thermal history of the LDED process governs both macro- and microstructures, potentially influencing the mechanical properties of deposited parts. Yang et al. [25] investigated the effect of interlayer cooling time on thermal accumulation effects, establishing the correlation between interlayer cooling time and forming quality. Scime et al. [26] measured multiple melt pool cross-sections to ensure the melt pool geometry. Qiu et al. [27] utilized cold metal transfer-assisted LDED to fabricate nickel-based Hastelloy C276 alloy, reducing mechanical anisotropy through interlayer temperature control. As shown in Fig. 1.3, the initial rapid growth of the melt pool was associated with slow heat dissipation effects without preheating and substrate cooling. Beyond the initial six layers, the maximum variation in melt pool volume remained below 15%. Silveira et al. [28] investigated the relationship between intrinsic heat treatment (IHT) cooling rates and martensitic steel phase transformations, confirming that thermal accumulation effects influence dislocation density in austenitic stainless steels. Without additional measures to mitigate heat accumulation, continuous LDED processes will cause elevated substrate temperatures and poor heat dissipation conditions in the melt pool. It will also cause the

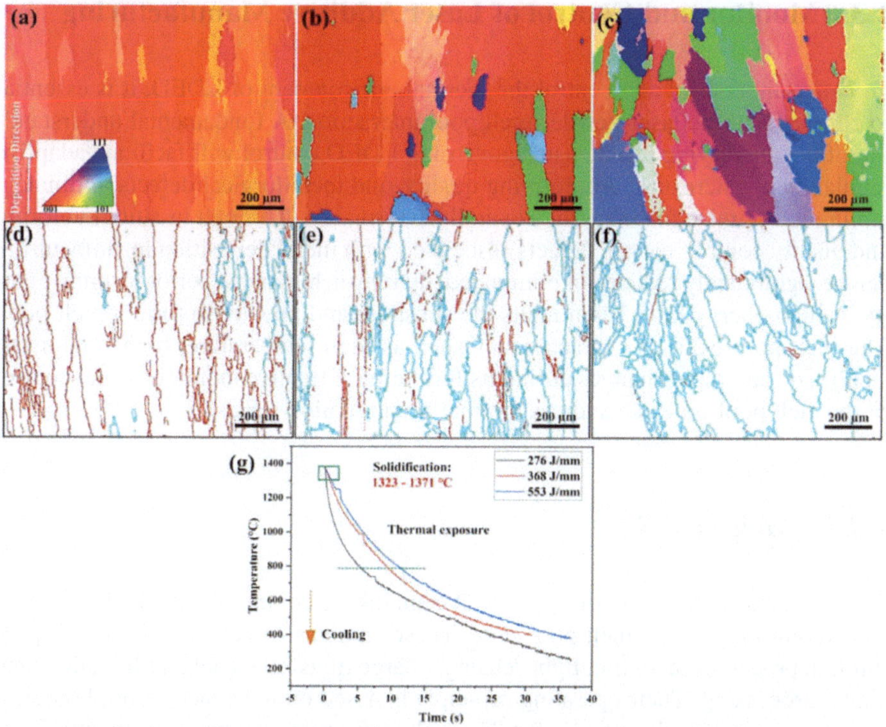

Fig. 1.3 EBSD orientation maps of the CMT-processed Hastelloy C276 samples fabricated under (**a**) 276 J/mm, (**b**) 368 J/mm, and (**c**) 553 J/mm heat inputs. (**d–f**) show the corresponding grain boundary maps of the above-mentioned three heat input conditions, with the LAGBs (misorientation 2–10°) and HAGBs (misorientation >10°) coloured in red and blue, respectively. (**g**) Temperature evolution during cooling of three samples

melt pool to grow in size along the deposition direction, leading to uneven track morphology or severe collapse issues [29, 30].

To address thermal effects, modifying laser process parameters often requires extensive exploratory orthogonal experiments and optical verification to determine suitable settings. This offline detection-verification approach not only increases time and material costs but also fails to address the root cause originating from the heat source. The thermal history during LDED encompasses heating rates, cooling rates, and temperature gradients. It significantly influences not only forming quality and microstructure but ultimately impacts the mechanical and chemical properties of the part [31].

1.3 Monitor and Control of Laser Additive Manufacturing

To better understand and control the thermal environment of LDED, it is essential to comprehend the laser-powder-melt pool interaction. A fundamental understanding of the underlying mechanisms governing LDED control will facilitate adaptive regulation of microstructure, forming quality, and metallurgical defects, ultimately aiming to optimize the material's overall performance. Through in-situ monitoring and quality control, certain defects associated with metal deposition manufacturing can be significantly reduced or eliminated, thereby achieving superior forming quality. A high-energy laser beam melts the substrate and powder to form a melt pool. The melt pool exhibits characteristics such as high temperature, high brightness, small size, and rapid changes during its formation. Effective monitoring and control of the melt pool state can address thermal accumulation issues.

1.3.1 Melt Pool Size

In current research methodologies, CCD cameras are commonly employed for real-time monitoring of the melt pool state. These cameras detect and measure objects through processes including light sensing, charge transfer, signal amplification, and signal processing. Their operating principle is based on the concepts of photoelectric conversion and charge transfer. Through appropriate voltage control and signal processing, high-quality images and accurate melt pool shape detection results can be obtained.

When employing CCD cameras for melt pool monitoring, common mounting configurations include coaxial placement, positioning above the pool at an angle, and off-axis placement. Coaxial mounting typically facilitates capturing top-view images of the molten pool for subsequent width and length measurements. Two CCD cameras positioned at 45° angles to the melt pool are used for shape capture. Off-axis placement is typically employed for detecting melt pool depth. Monitoring melt pools with CCD cameras offers the following advantages: (1) Flexible positioning based on specific shape detection requirements; (2) High detection accuracy and rapid efficiency; (3) Dynamic melt pool shape monitoring provides technical assurance for subsequent deposition quality control.

As shown in Fig. 1.4, Lane et al. [32] provided an example of using dynamic laser absorption measurements combined with CCD camera monitoring to inform the interpretation of signals for physical properties and process monitoring. Heralica et al. [33] employed camera monitoring and iterative learning to control the width and height of deposited layers by adjusting the wire feed rate for each layer. Akbari et al. [34] utilized a CMOS coaxial camera to monitor melt pool geometry, achieving deposition components with favorable cooling rates. They compared microstructures and tensile properties between thin-walled and block components. The results demonstrated that real-time monitoring of the melt pool enables control over

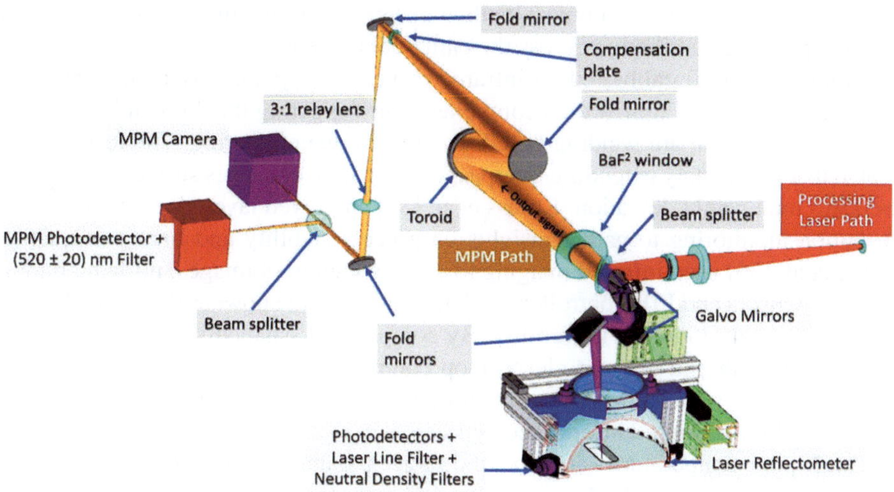

Fig. 1.4 Computer model of the heating laser path (from right), the reflectometer (bottom), and the co-axial melt pool monitoring (MPM) path

the solidification microstructure of deposited components. Previous studies indicated that optical camera monitoring of melt pool state or shape detection lays the foundation for subsequent adjustments to laser process parameters to enhance the forming quality. Monitoring melt pool state is the first step in achieving component quality control, making high-precision, high-efficiency monitoring of molten pool state critically important.

1.3.2 Melt Pool Temperature

During laser additive manufacturing, the instantaneous temperature of the melt pool, its temperature gradient, and the accumulated temperature at its surface all influence the deposition quality of the component. The melt pool temperature can be monitored using infrared thermal imaging equipment. The temperature measurement principle is based on the law of thermal radiation: the temperature of the molten pool is proportional to its thermal radiation intensity. The infrared thermal imager captures the infrared radiation emitted by the molten pool and converts it into an image, from which the temperature distribution of the melt pool can be obtained. It is important to note that due to the extremely high temperature of the melt pool, the lens of the thermal imager should not be positioned too close to the laser beam, as this could damage the imager. By plotting the temperature curve throughout the entire deposition process, it is possible not only to monitor the melt pool temperature at each moment but also to calculate the temperature gradient and cooling rate at each instant. Infrared thermal imaging for melt pool temperature detection is a passive, non-contact method that provides an intuitive visualization of the temperature distribution across the entire component.

Yang et al. [25] calibrated an infrared camera to capture surface temperatures during deposition of thin-walled components, investigating the thermal behavior of gas-shielded metal arc welding based on additive manufacturing. The results indicated that moderately extending interpass cooling time enhances surface quality and reduces total height deviation. Krauss et al. [35] employed layered temperature distribution monitoring to gather insights on process stability and final part quality. Ding et al. [36] used infrared imaging to capture melt pool images, utilizing photoelectric sensors and PID controllers to detect and regulate powder flow rate, thereby enhancing melt pool size uniformity. By comparing L-shaped single-wall components built with and without closed-loop control, they demonstrated that components prepared with melt pool temperature monitoring combined with closed-loop control exhibited more uniform geometries.

Melt pool temperature serves as the target parameter for feedback control of thermal effects during LDED and is a key indicator for characterizing thermal accumulation effects in thin-walled parts. As a critical factor determining component forming quality, melt pool temperature and size influence not only powder utilization but also the uniformity of macro-morphology and microstructure. High-precision detection of melt pool temperature lays the foundation for subsequent melt pool temperature control.

1.3.3 Quality Control

Quality control in laser additive manufacturing is crucial for ensuring component quality and production efficiency. Quality control holds the following significance: (1) By controlling material quality, optimizing process parameters, and implementing real-time monitoring, it effectively reduces product defects and deformation while enhancing surface precision and forming quality; (2) Quality control lowers manufacturing costs and risks, timely detection and resolution of quality issues minimize scrap rates and rework rates; (3) Quality control further enhances corporate competitiveness, as enterprises with high-quality products more readily gain market trust and recognition.

Common quality control methods include: (1) Selecting appropriate materials prior to manufacturing, as those used in the additive process significantly impact final product quality; (2) Controlling external factors in the manufacturing environment—such as gas composition, humidity, and pressure during deposition—to improve product quality; (3) Adjusting process parameters—such as laser power, scanning strategy, scanning speed, and powder feed rate—can reduce deposition defects and deformation; (4) Real-time detection of defect types in deposited components enables integrated additive/subtractive quality control through processes like milling removal; (5) Real-time monitoring of component formation during additive manufacturing, coupled with feedback control systems, achieves online quality control and effectively enhances forming efficiency.

To address component quality issues, altering laser process parameters often requires extensive exploratory orthogonal experiments and optical validation to determine optimal settings. This offline inspection-verification approach not only increases time and material costs but also fails to address the root cause from the heat source perspective. Furthermore, while online defect detection and integrated subtractive control for deposited component quality can detect and remove internal defects, it cannot control surface quality and forming accuracy, additionally increasing time and material costs.

1.4 Scope of the Book

Addressing the characteristics of conventional laser directed energy deposition, a novel method called "melt pool in-situ controlling of laser additive manufacturing technology" by integrating melt pool morphology and temperature gradient regulation was proposed. This technique monitors melt pool morphology to reflect temperature gradient variations, utilizing real-time laser power adjustment to mitigate thermal accumulation effects during thin-walled component deposition. This enhances cooling rates to achieve microstructural customization. Closed-loop feedback of melt pool dimensions improves dimensional accuracy and surface quality of thin-walled components, further enhancing mechanical properties.

The core principle of this technology involves monitoring thermal history during deposition—including temperature gradients, heating rates, and cooling rates—and integrating process parameter control to achieve stable melt pool development. This research holds significant theoretical and practical value, representing a key approach to forming quality control in laser additive manufacturing.

This book details the application of in-situ melt pool control-assisted laser directed energy deposition in high-performance manufacturing of critical components. Metal additive manufacturing (AM) technologies have made considerable progress in the basic theoretical field since its invention in the 1970s. However, challenges persist in the synergistic control of forming quality and performance, primarily involving dimensional accuracy and microstructure regulation issues caused by cumulative thermal effects. This book systematically compares traditional LDED and LDED with in-situ melt pool control in terms of surface quality, microstructure, and overall performance for single-track, single-layer, thin-walled 316L components. It highlights core research findings, laying the theoretical foundation for elucidating the principles and effects of thermo-mechanical interactions in laser metal additive manufacturing. The book is divided into four parts, comprehensively presenting laser metal additive manufacturing technology based on thermo-mechanical interactions. The first part reviews the background of melt pool in-situ control technology in laser metal additive manufacturing. The second part elaborates on the theoretical foundations of LDED and closed-loop control technology. The remaining two parts specifically describe the microstructural evolution, forming quality, and overall performance of materials when employing in-situ melt pool control technology.

The book outline is as follows:

In Chap. 2, the authors focus on the equipment configuration, monitoring methods, and control theory for in-situ melt pool control.

In Chap. 3, the authors investigate the effects of in-situ melt pool control on the microstructure of single-track and single-layer laser directed energy deposition (LDED) of 316L stainless steel, examining aspects such as melt pool dilution rate, surface morphology, metallurgical defects, and microstructure.

In Chap. 4, the authors discuss the microhardness, wear resistance, and tensile properties of 316L SS coatings deposited using in-situ melt pool control technology. It explores the mechanisms by which microstructural evolution influences coating wear performance and strength-toughness balance.

In Chap. 5, the authors derive thermal effect models for both deposition modes, focusing on interlayer microstructural differences in 316L SS planar thin-walled components. It analyzes the influence of melt pool thermal history and cooling rates on columnar dendrite recrystallization evolution.

In Chap. 6, the authors focus on analyzing the forming quality of 316L SS planar thin-walled components under both deposition modes. Electrochemical corrosion and tensile tests were conducted to elucidate the influence mechanism of in-situ melt pool control technology on corrosion resistance and toughness.

In Chap. 7, the authors fabricate 316L SS curved thin-walled components with varying curvatures and power levels under both deposition modes, characterizing their interlayer microstructure and forming quality.

In Chap. 8, a thermal effect model for 316L SS curved thin-walled components was derived. Tensile property tests revealed a unique quasi-superplastic phenomenon.

References

1. Zhu J Q, Ling Z M, Du F R, Ding X P and Li H M 2018 Monitoring of laser metal-wire additive manufacturing temperature field using infrared thermographyInfrared Laser Eng. 47 0604002.
2. P. Spittle. Gas turbine technology, Special Feature: Flight, 2003, 38 (6): 504-511.
3. T. Wohlers, T. Caffrey, Wohlers Report 2013- Annual Worldwide Progress Report of Additive Manufacturing and 3D Printing State of the Industry. USA: Wohlers Associates, 2019.
4. X.D. Nong, X.L. Zhou, Y.D. Wang, et al. Effects of geometry, location, and direction on microstructure and mechanical properties of 15–5PH stainless steel fabricated by directed energy deposition. Materials Science and Engineering: A, 2021, 821: 141587.
5. J.Z. Lu, H.F. Lu, X. Xu, et al. High-performance integrated additive manufacturing with laser shock peening –induced microstructural evolution and improvement in mechanical properties of Ti6Al4V alloy components. International Journal of Machine Tools & Manufacture, 2020, 148: 103475.

6. L. R. Goossens, B. V. Hooreweder, A virtual sensing approach for monitoring melt-pool dimensions using high speed coaxial imaging during laser powder bed fusion of metals, Addit. Manuf. 40 (2021) 101923, https://doi.org/10.1016/j.addma.2021.101923.
7. G. Mohr, S. J. Altenburg, A. Ulbricht, P. Heinrich, D. Baum, C. Maierhofer and K. Hilgenberg, 'In-situ defect detection in laser powder bed fusion by using thermography and optical tomography—comparison to computed tomography," Metals, vol. 10 (103), no. 1, 2020.
8. X. P. Li et al., Selective laser melting of nano-TiB2 decorated AlSi10Mg alloy with high fracture strength and ductility. Acta Mater 129, 183–193 (2017). https://doi.org/10.1016/j.actamat.2017.02.062.
9. C. T. Kwok, H. C. Man, F. T. Cheng, K. H. Lo, Developments in laser-based surface engineering processes: With particular reference to protection against cavitation erosion. Surf. Coat. Technol. 291, 189–204 (2016). https://doi.org/10.1016/j.surfcoat.2016.02.019
10. Hull Charles W. Chuck hull invents stereolithography or 3D printing and produces the first commercial 3D printer.
11. Becker D, Boley S, Eisseler R, Stehle T, Möhring H C, Onuseit V, Hoßfeld M and Graf T. 2021. Influence of a closed-loop controlled laser metal wire deposition process of S Al 5356 on the quality of manufactured parts before and after subsequent machining. Prod. Eng. Res. Dev. 15, 489–507.
12. S.H. Liao, S. Webster, D. Huang, R. Council, K. Ehmann, J. Cao. Simulation-guided variable laser power design for melt pool depth control in directed energy deposition, Addit. Manuf. 56 (2022) 102912, https://doi.org/10.1016/j.addma.2022.102912.
13. P. A. Rometsch, Y.M. Zhu, X.H. Wu, et al. Review of high-strength aluminium alloys for additive manufacturing by laser powder bed fusion. Materials & Design, 2022, 219: 110779.
14. Kim, S., Jeon, I., & Sohn, H. (2023). Infrared thermographic imaging based real-time layer height estimation during directed energy deposition. Optics and Lasers in Engineering, 168, 107661. https://doi.org/10.1016/j.optlaseng.2023.107661.
15. A. Emelogu, M. Marufuzzaman, S.M. Thompson, et al. Additive manufacturing of biomedical implants: a feasibility assessment via supply-chain cost analysis. Additive Manufacturing, 2016, 11: 97–113.
16. J.F. Sun, H.F. Lu, Z. Wang, et al. High-temperature oxidation behaviour of Ti65 titanium alloy fabricated by laser direct energy deposition. Corrosion Science, 2024, 229:111866.
17. W. Fan, Y.J. Peng, Y. Qi, et al. Partially melted powder in laser based directed energy deposition: Formation mechanism and its influence on microstructure. International Journal of Machine Tools & Manufacture, 2023, 192: 104072.
18. Z. Sun, W. Guo, L. Li. In-process measurement of melt pool cross-sectional geometry and grain orientation in a laser directed energy deposition additive manufacturing process. Optics & Laser Technology, 2020, 129: 106280.
19. L. Sinclair, S.J. Clark, Y.H. Chen, et al. Sinter formation during directed energy deposition of titanium alloy powders. International Journal of Machine Tools & Manufacture, 2022, 176: 103887.
20. M. H. Farshidianfar, A. Khajepour, A. P. Gerlich. Real-time control of microstructure in laser additive manufacturing. The International Journal of Advanced Manufacturing Technology, 2015, 82: 1173–1186.
21. M. H. Farshidianfar, A. Khajepour, A. P. Gerlich. Effect of real-time cooling rate on microstructure in Laser Additive Manufacturing. The International Journal of Advanced Manufacturing Technology, 2016, 231: 468-478.
22. S.H. Li, B. Chen, C.W. Tan, et al. Effects of oxygen content on microstructure and mechanical properties of 18Ni300 maraging steel manufactured by laser directed energy deposition. Optics & Laser Technology, 2022, 153: 108281.
23. X. Xu, H.F. Lu, Y.Y. Su, et al. Comparing corrosion behavior of additively manufactured Cr-rich stainlesssteel coating between conventional and extreme high-speed laser metal deposition. Corrosion Science, 2022, 195: 109976.

24. Y.Y. Su, Z.F. Wang, J.C. Xie, et al. Microstructures and mechanical properties of laser melting deposited Ti6Al4V/316L functional gradient materials. Materials Science and Engineering: A, 2021, 817: 141355.
25. D.Q. Yang, G. Wang, G.J. Zhang. Thermal analysis for single-pass multi-layer GMAW based additive manufacturing using infrared thermography. Journal of Materials Processing Technology, 2017, 244: 215–224.
26. L. Scime, J. Beuth. Melt pool geometry and morphology variability for the Inconel 718 alloy in a laser powder bed fusion additive manufacturing process. Additive Manufacturing, 2019, 29: 100830.
27. Z.J. Qiu, Z.Y. Wang, A.A. Gazder, et al. Stabilised mechanical properties in Ni-based Hastelloy C276 alloy by additive manufacturing under different heat inputs incorporated with active interlayer temperature control. Materials Science and Engineering: A, 2023, 862: 144434.
28. A. C. F. Silveira, R. Fechte-Heinen, J. Epp. Microstructure evolution during laser-directed energy deposition of tool steel by in situ synchrotron X-ray diffraction. Additive Manufacturing, 2023, 63: 103408.
29. Montevecchi F, Venturini G, Grossi N, et al. Idle time selection for wire-arc additive manufacturing: A finite element-based technique. Additive Manufacturing, 2018, 21:479–86.
30. Ding D, Pan Z, Cuiuri D, et al. Wire-feed additive manufacturing of metal components: technologies, developments and future interests. The International Journal of Advanced Manufacturing Technology, 2015, 81: 465–81.
31. S.M. Thompson, L.K. Bian, N. Shamsaei, et al. An overview of Direct Laser Deposition for additive manufacturing; Part II: Mechanical behavior, process parameter optimization and control. Additive Manufacturing, 2015, 8: 12–35.
32. B. Lane, I. Zhirnov, S. Mekhontsev, et al. Transient laser energy absorption, co-axial melt pool monitoring, and relationship to melt pool morphology. Additive Manufacturing, 2020, 36: 101504.
33. A. Heralica, A. Christianssona, B. Lennartson. Height control of laser metal-wire deposition based on iterative learning control and 3D scanning. Optics and Lasers in Engineering, 2012, 50: 1230–1241.
34. M. Akbari, R. Kovacevic. An investigation on mechanical and microstructural properties of 316LSi parts fabricated by a robotized laser/wire direct metal deposition system. Additive Manufacturing, 2018, 23: 487-497.
35. H. Krauss, T. Zeugner, M.F. Zaeh. Layerwise monitoring of the selective laser melting process by thermography. Physics Procedia, 2014, 56: 64–71.
36. Y.Y. Ding, J. Warton, R. Kovacevic. Development of sensing and control system for robotized laser-based direct metal addition system. Additive Manufacturing, 2016, 10: 24-35

Chapter 2
Melt Pools In-Situ Control Technology-Assisted Laser Directed Energy Deposition

2.1 Introduction

There are many characteristics of the melt pool, such as high temperature, high brightness, small size, and rapid changes [1–3]. When different processing parameters were set, the shape, width, depth, and fluidity of the melt pool all affect the deposition quality of the component. Therefore, online monitoring of the melt pool is critical during laser deposition processing. Additionally, compared to post-processing methods for improving forming quality, real-time monitoring of the melt pool and online data collection can effectively achieve the melt pool state and thereby enhance deposition quality. Therefore, establishing a melt pool online detection and control system can reduce manufacturing costs and improve production quality [4–6].

2.2 Laser Directed Energy Deposition Equipment

2.2.1 Equipment Composition

Figure 2.1 shows the ILAM®20SrE-600 in-situ melt pool control-assisted laser directed energy deposition equipment. The equipment integrates key components such as a Laserline fiber laser with a maximum power of 6000 W, a robotic arm and its controller, a dual-tube powder feeder, a CCD optical imaging system, a laser water cooler, a two-axis rotary table, a PLC control system, and a glove box. The fiber laser model is LDF6000-60, with a wavelength range of 900–1100 nm and a beam parameter product (BPP) of 66 mmmrad. The robotic arm model is RX160, capable of six-axis. Argon is used as the inert gas and protective gas, with a flow rate

K. Luo et al., *Melt Pool In-Situ Controlling of Laser Additive Manufacturing Technology*, https://doi.org/10.1007/978-981-95-4597-1_2

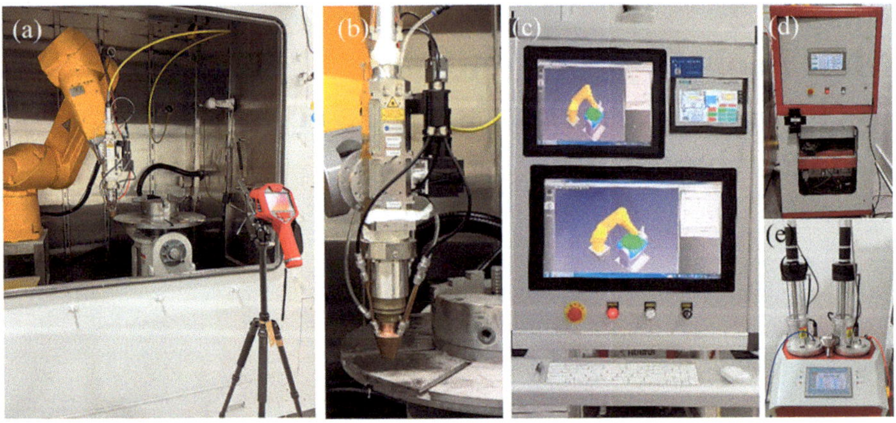

Fig. 2.1 (**a**) Melt pool in-situ control-assisted laser directed energy deposition equipment, (**b**) coaxially mounted CCD camera and laser head; (**c**) melt pool monitoring and control PC terminal; (**d**) laser line, (**e**) powder feeder

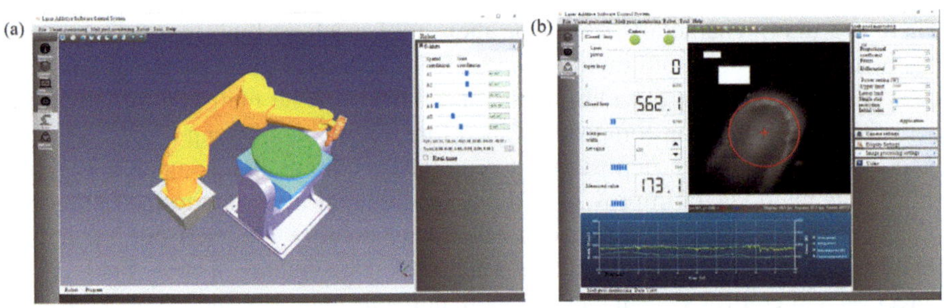

Fig. 2.2 (**a**) Computer-aided Additive Manufacturing (CAD) software system interface; (**b**) closed-loop feedback control system interface

of 20 L/min. To ensure uniform powder flow and deposition efficiency, a coaxial three-point nozzle is employed. The laser beam forms a spot with a diameter of 3 mm below the focal point. The specific composition and functions of the optical detection device are detailed in Sect. 2.3.2.

The system was developed in conjunction with a computer-aided additive manufacturing and closed-loop feedback technology system (Fig. 2.2). The computer-aided additive manufacturing software system primarily includes functions such as component slicing, path planning, and parameter settings, enhancing the system's operability and reliability in industrial environments. The closed-loop feedback control system can monitor parameters such as laser spot size, melt pool shape, and temperature gradients in real time using optical or thermal sensors, and provide closed-loop feedback signals to the control system, significantly improving the manufacturing accuracy and the reliability of laser direct deposition.

2.2.2 *Materials*

316L stainless steel has the advantages of high hardness, high wear resistance, and high toughness [7, 8], making it an important material for manufacturing large-scale engineering components [9, 10]. However, due to prolonged exposure to harsh and complex working environments, the surfaces of components are prone to failure and scrapping, resulting in significant material waste and high costs. The substrate and powder used in the deposition experiments in this study were both 316L stainless steel, with their chemical compositions shown in Table 2.1. Prior to the experiments, the substrate was ground and polished using a grinding wheel, and the polished surface was cleaned with an ethanol and acetone solution to remove residual oil and impurities from the substrate surface. The 316L powder selected for deposition was prepared using the gas-phase atomization method, and dried in a drying oven. The particle size range of the 316L powder is 53–150 μm.

2.2.3 *Forming Technology*

The process flowchart of the laser-directed energy deposition assisted by melt pool in-situ control is shown in Fig. 2.3. 316L stainless steel powder is transported from the powder feeder through the powder supply pipe to the nozzle, where it is melted by a high-energy laser beam to form a melt pool. An optical imaging system, specifically a CCD camera, is coaxially mounted with the laser head and moves in sync with it. This system primarily captures images of the top surface of the melt pool during the forming process and extracts the melt pool width. The obtained melt pool width values are then plotted as a continuous dynamic curve. To obtain a stable melt pool width, the PID controller integrated into the control system adjusts the laser power to suppress the characteristic of the melt pool width continuously increasing. Based on the melt pool in-situ control process, this paper has deposited various shapes, including single-track, single-layer, thin-walled parts, and curved thin-walled parts.

Table 2.1 Chemical compositions of 316L powder (wt%)

Elements	C	P	Mn	Si	Mo	Ni	Cr	Fe
Content	0.006	0.027	0.764	0.35	2.21	10.46	16.65	balanced

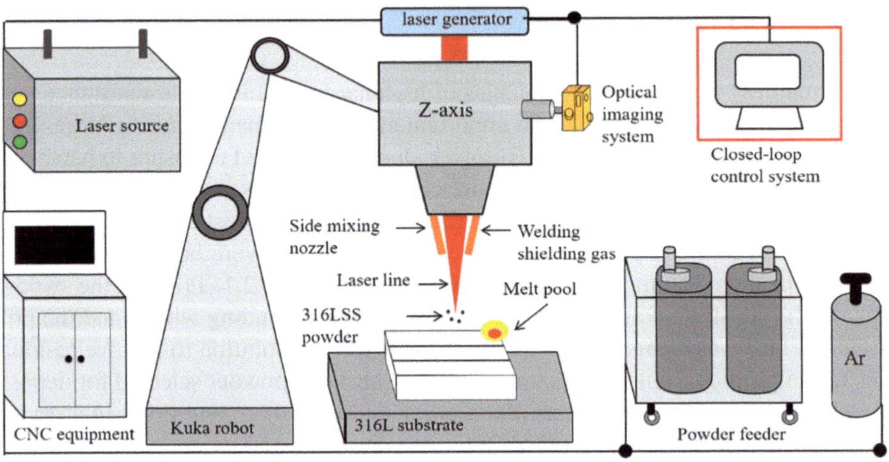

Fig. 2.3 Mechanism diagram of the in-situ molten pool control-assisted LDED process

2.3 Melt Pool Online Monitor

2.3.1 Characteristics of the Melt Pool Image

Combining the processing principles and characteristics of laser-directed energy deposition, the high-energy laser beam from the laser head is directed onto the substrate, melting the powder in the powder feeder and metallurgically bonding it with the substrate material to process the component. During this process, the substrate rapidly heats up, melting the metal powder to form a melt pool. The melt pool typically has the following characteristics [11–18]:

1. High temperature: The powder materials used in laser deposition are typically titanium alloys, aluminum alloys, stainless steel, etc., with melting points exceeding thousands of degrees. The temperature of the melt pool formed after the metal powder melts is extremely high, necessitating careful selection of measurement equipment and working distance;
2. High brightness: Under high temperatures, the melt pool emits intense thermal radiation to the surrounding environment, producing blinding light that cannot be directly observed with the naked eye. This imposes requirements on the imaging principles of the melt pool detection system;
3. Small size: Due to the extremely small spot diameter of the laser beam (typically 3–4 mm), the melt pool area is very small, necessitating high precision and high resolution in the detection system;
4. Rapid changes: In actual laser deposition processes, to ensure deposition efficiency and production progress, relatively high laser scanning speeds are typically employed. However, the shape of the melt pool formed after powder

melting is highly variable, so the acquisition frequency of the acquisition system must be maintained.

2.3.2 Melt Pool Image Processing and Width Extraction

The melt pool monitoring system typically includes the following modules: camera control module, image acquisition module, image processing module, image display module, process parameter control b module, real-time data display module, data storage module, and video playback module. The relationships between each module are shown in Fig. 2.4a.

The accuracy and timeliness of melt pool image acquisition are the first step in the melt pool monitoring and control process. Integrating the optical signal acquisition and monitoring device for the melt pool state with the laser head not only ensures an appropriate working distance and protects the equipment but also enables coaxial movement with the laser head, ensuring the real-time and accurate acquisition of data [19, 20]. Additionally, to obtain clear imaging, an optical detection device with sufficient resolution and signal-to-noise ratio should be employed. As shown in Fig. 2.4b, the hardware components of the melt pool state acquisition system include the following devices: network cable, Daheng MER-E-P industrial camera, coaxial imaging lens, 45° reflector, coaxial illumination LED light source with collimator, 45° semi-transparent/semi-reflective beam splitter, main beam splitter for laser and visible light, filter, and LED light source power cable. During the laser deposition process, the radiation from the melt pool is generally reflected back through the optical path in the laser head to form an image of the melt pool. The radiation from the melt pool passes through the filter and lens into the CCD

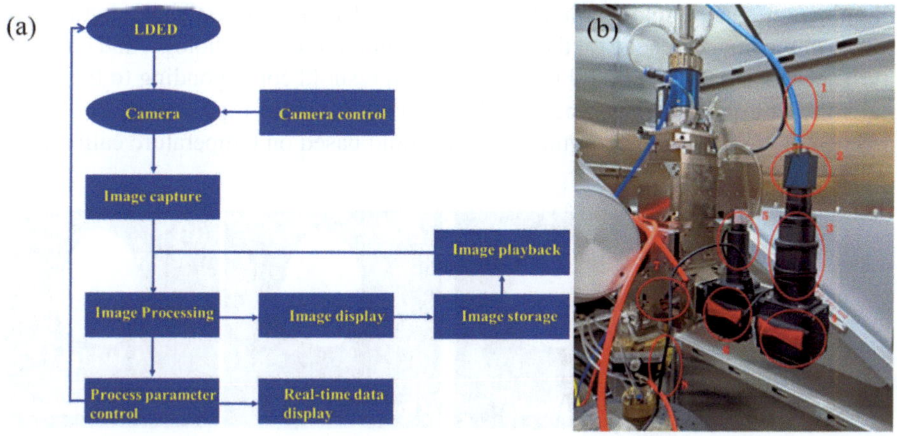

Fig. 2.4 (**a**) Flow chart for online monitoring of melt pool; (**b**) optical device for online monitoring of the melt pool

camera, ultimately forming an image of the melt pool. The coaxial imaging lens is equipped with an aperture and focus adjustment to regulate the amount of light passing through the lens, thereby achieving appropriate brightness and a clear image. The collimating mirror primarily allows more light to pass through the laser head optical path, facilitating its return along the same path after exiting through the small aperture of the nozzle. The camera's acquisition frequency and resolution are 50 fps and 64 × 48 pixels, respectively.

Based on the image signals captured by the CCD camera, it is crucial to employ suitable and efficient image processing algorithms to extract melt pool feature information in real-time [21–23]. As shown in Fig. 2.5, the basic steps of melt pool image processing generally include mean filtering for noise reduction, threshold segmentation, and contour smoothing. During the acquisition of melt pool images, image noise is generated due to interference from random signals, which is a multidimensional random process [24, 25]. This noise is typically analyzed using probability distribution functions and probability density distribution functions. The mean filtering is employed to process melt pool images. First, the grayscale values of image pixels are recorded in an array. Then, the radius value of the box is set, and the average of all pixel values within the box is calculated. The resulting value represents the grayscale value of the corresponding pixel after mean filtering.

As shown in Fig. 2.5b, to extract the melt pool contour and simplify the subsequent width extraction process, a binary threshold segmentation method is employed. This method is generally suitable for situations where the target and background have strong contrast. By setting the threshold T, the image is divided into two main parts: pixels greater than T and pixels less than T. Threshold setting is the most critical parameter in the image processing algorithm for melt pool width extraction, used to distinguish the melt pool from the background. Specifically, the largest connected region with gray values greater than the threshold represents the molten pool, while the remainder constitutes the background. Under normal conditions, the larger the threshold, the smaller the calculated melt pool width. In the absence of illumination, the melt pool emits light through self-radiation, so the pixel grayscale values in the image depend on the material's temperature, morphology, and emissivity. Theoretically, there should be a grayscale threshold corresponding to the material's melting point that can accurately separate the solid-liquid interface, but in practice, it is difficult to determine this threshold based on temperature calibration.

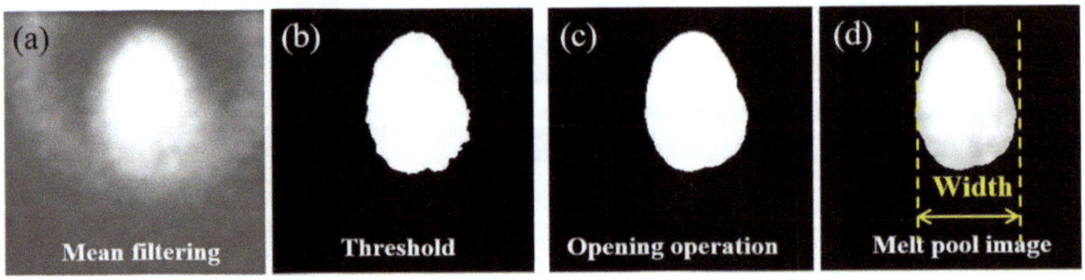

Fig. 2.5 Melt pool image processing steps (**a**) Mean filtering, (**b**) threshold, (**c**) opening operation, (**d**) melt pool image

From an image processing perspective, histogram analysis can be performed on the image to find a threshold that separates the foreground from the background. Opening operations are typically used to eliminate small objects or separate contacting objects, smooth object boundaries, and maintain their area without significant alteration, and can be applied to smooth the melt pool contour.

In actual deposition components, to obtain a top-down view of the melt pool, the optical path of the CCD camera is integrated into the deposition head. As shown in Fig. 2.6, first, the melt pool image is corrected. Two directions are identified within the field of view of the coaxial camera, and images of the deposited thin-walled component are captured in a cross-scan mode to obtain a complete melt pool image. Based on the significant difference in gray values between the molten pool and other regions, the triangular threshold segmentation method is used to determine the melt pool region. Additionally, during morphological processing, the contour of the region of interest is subjected to opening and smoothing operations to obtain the melt pool width image. Since the direction of melt pool movement is random, the inscribed circle algorithm is used to extract the melt pool width. Finally, the actual width of the melt pool is calculated using pixel values and the working distance.

2.3.3 Temperature Collection of the Melt Pool

To detect the temperature gradient and cumulative temperature of the melt pool and characterize its thermal history data, an infrared thermal imaging camera is generally used for online monitoring [26–28]. Its working principle involves utilizing the infrared radiation energy distribution image emitted by the melt pool to determine the temperature distribution of the measured object, thereby achieving the purpose of infrared temperature measurement. When the melt pool absorbs or reflects infrared radiation energy, corresponding electrical signals of varying strength are generated in the detector. These signals are amplified by an amplifier and shaping circuit before being input into an integrator for integration processing. After processing by a microprocessor, a thermocouple outputting a temperature value corresponding to the target is generated. The infrared thermal imaging camera is equipped with overheat protection and overcurrent protection devices to ensure safety and reliability.

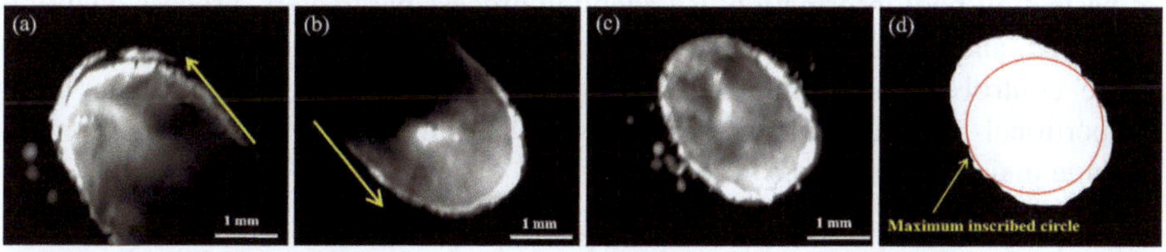

Fig 2.6 (**a, b**) Images of the melt pool trajectory under both scanning directions, (**c**) image of the complete melt pool, (**d**) inline-circle algorithm used to extract the width of the melt pool

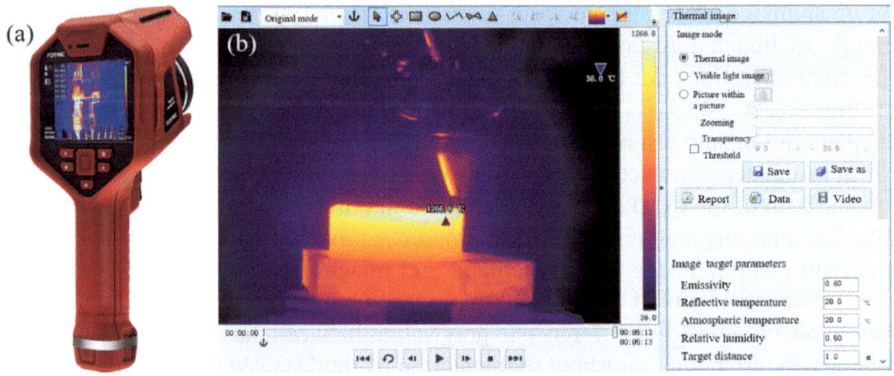

Fig. 2.7 (**a**) Infrared thermal imaging camera, (**b**) thermal imaging camera workspace interface

As shown in Fig. 2.7, this paper uses the FOTRIC340+ model infrared thermal imaging camera with a separately mounted off-axis sensor. It can achieve laser, image automatic, and continuous focusing, with a temperature measurement range of −20 °C to 1550 °C, enabling precise measurement of the solid-liquid phase temperature of metallic materials. The infrared resolution reaches 640 × 480 pixels, and the thermal sensitivity is 30 mK (0.03 °C).

2.4 Melt Pool In-Situ Control

2.4.1 PID Control Strategy

Melt pool in-situ control generally refers to methods for monitoring the state of the melt pool and controlling its width in real time during LDED [29–32]. Therefore, online monitoring and extraction of the melt pool width is the first step in an in-situ melt pool control system. To improve the quality of component formation, correcting unexpected melt pool widths and maintaining their stability are key factors in evaluating the effectiveness of in-situ melt pool control. In actual industrial production, PID control algorithms are commonly used for closed-loop feedback control of basic melt pool information. By adapting process parameters, process quality is improved, and process stability is enhanced. PID control is a widely applied automatic control method in industrial control systems, with its full name being Proportional-Integral-Derivative Control (PID Control). A PID controller consists of three main components:

1. Proportional control: Adjusts the output in proportion to the error magnitude to bring the controlled object's output closer to the desired value;

2. Integral control: Adjusts the output based on the cumulative error to eliminate static errors in the system and enable the output to more accurately track the setpoint;
3. Derivative control: Adjusts the output based on the rate of change of the error to predict future error trends, enabling proactive control to avoid overshoot or oscillation and accelerate the system's response speed.

The PID controller forms the control deviation e(t) based on the setpoint and actual output values, and combines the deviation proportionally, integrally, and derivatively through a linear combination to form the control signal for controlling the controlled object [33, 34].

The key point to PID control lies in the reasonable balance of the control functions for three parts to meet the control requirements of the controlled object. In practical applications, the performance of a PID controller can be evaluated and measured by indicators such as response time, overshoot, and steady-state error. PID control has the advantages of simple structure, good robustness and high reliability. However, since actual objects usually have characteristics such as nonlinearity, time-varying uncertainty and strong interference, it is necessary to determine appropriate parameters through PID parameter tuning to ensure the stability of the control system.

In the melt pool in-situ control system, process parameters are pre-set and maintained at constant values. Due to the thermal accumulation effect of continuous deposition, even when the laser power remains constant, the actual thermal input value of the melt pool exceeds the actual laser power value, leading to a continuous increase in melt pool width. To address the issue of poor forming quality caused by the continuous increase in melt pool width due to the thermal accumulation effect, the laser power should be adjusted online to correct the uniformity and stability of the melt pool width.

Currently, there are no specific standards to define the ideal value for melt pool width in deposition processes, so there is no reference basis for setting the ideal value. Since the thermal accumulation effect is relatively small during the early stages of deposition, the average width of the first three to five layers of each deposition is taken as the ideal width. The difference between the detected real-time melt pool width and the set ideal melt pool width is used as the input to the controller, where represents the ideal value of the set melt pool width, and represents the real-time detected melt pool width. The output of the entire controller is the change in laser power, denoted as

$$\Delta P = K_P \left\{ e(t) + \frac{1}{T_I} \sum_{t}^{m=1} e(m) + T_D \left[e(t) - e(t-1) \right] \right\} \tag{2.1}$$

In addition, in order to suppress the growth of the melt pool width and ensure the consistency of the melt pool width, the change rate of the melt pool width detected in real time is set as

$$\sigma = \frac{W_{\mathrm{r}}(k) - W_{\mathrm{i}}(k)}{W_i(k)} \tag{2.2}$$

when the change in melt width is controlled within ±3%, it is considered to meet the stability requirements of the melt pool width.

2.4.2 PID Parameter Setting

PID control plays a crucial role in the feedback control of the melt pool width, where the selection of the three parameters: KP, KI, and KD, directly impacts the stability and effectiveness of the entire controller. The tuning of PID parameters directly affects the system's stability. If tuned improperly, it can lead to slow response times, large overshoot, prolonged regulation times, and even system oscillation or instability. Generally, when the actual value oscillates near the target value, KI should be increased. If oscillation persists, KP should be appropriately reduced. If the deviation between the actual value and the target value is significant, KP should be increased, KI reduced, or KD increased. A low-pass filter is used to filter out most noise, thereby eliminating interference with derivative control. The values of KP, KI, and KD determine the fluctuations in the curve. If these three parameters are improperly selected, the following three situations may occur: (1) overshoot due to excessive proportional gain or short integral time; (2) increased settling time due to insufficient proportional gain or long derivative time; (3) difficulty in reaching the setpoint due to insufficient proportional gain or long integral time.

In this paper, the determination of PID control parameters is obtained through engineering tuning methods. In actual laser deposition melt pool in-situ control systems, due to differences in the shape and power of deposited components, parameter optimization should be adjusted in real-time based on actual conditions. Generally, the proportional parameter KP is first set, initially adjusted to a larger value and then gradually reduced for correction, as adjusting KP can quickly bring the response curve close to the desired setpoint line. Next, the values of KI and KD are set to reduce oscillations and errors in the response melt pool width curve. The specific steps are as follows: first adjust KP, set KI and KD to 0, and continue until the melt pool width curve stabilizes after three step responses. KP is generally set to 0.7 times the value of the first step response. Next, adjust KI until system oscillations are effectively slowed down, at which point the appropriate KI value is achieved. Finally, the KD value is generally set to 0 or 0.05. In summary, PID parameters must be determined based on the fluctuation range of the standard melt pool width, and adjusting the sensor gain can increase signal strength. During the initial stage of the control mode, KP accelerates the system response while significantly suppressing the melt pool width. Therefore, KI and KD are enhanced to effectively reduce width suppression and slow down oscillations. Additionally, this method improves control stability before adjusting the melt pool to the target width. During the deposition process, the three-component parameters are adjusted based on the fluctuation range of the standard melt pool width.

2.5 Summary

This chapter systematically introduces the melt pool in-situ control system and equipment, including the principles of melt pool online monitoring, system composition, image processing procedures, control effect display, as well as the PID control strategy for online regulation of melt pool width and parameter tuning methods. Additionally, it describes the application process of melt pool online monitoring and in-situ control in laser directed energy deposition systems, as well as their impact on forming quality. It also covers the equipment composition of melt pool in-situ control assisting laser directed energy deposition processes, melt pool image processing algorithms, and closed-loop control strategies.

References

1. Y. Huang, T.G. Fleming, S.J. Clark, S. Marussi, K. Fezzaa, J. Thiyagalingam, C.L. A. Leung, P.D. Lee, Keyhole fluctuation and pore formation mechanisms during laser powder bed fusion additive manufacturing, Nat. Commun. 13 (1) (2022) 1170.
2. A.A. Martin, N.P. Calta, S.A. Khairallah, J. Wang, P.J. Depond, A.Y. Fong, V. Thampy, G.M. Guss, A.M. Kiss, K.H. Stone, C.J. Tassone, J. Nelson Weker, M. F. Toney, T. van Buuren, M.J. Matthews, Dynamics of pore formation during laser powder bed fusion additive manufacturing, Nat. Commun. 10 (1) (2019) 1987.
3. R. McCann, M.A. Obeidi, C. Hughes, E. McCarthy, D.S. Egan, R.K. Vijayaraghavan, A.M. Joshi, V. Acinas Garzon, D.P. Dowling, P.J. McNally, D. Brabazon, In-situ sensing, process monitoring and machine control in laser powder bed, fusion: A review. Addit. Manuf. 45 (2021).102058
4. H.P. Li, L. Hu, J.H. Ye, W. Wei, X.Y. Gao, Z. Qian, Y. Long, A high-precision in-situ monitoring system for laser directed energy deposition melt pool 3D morphology based on deep learning, Journal of Intelligent Manufacturing. https://doi.org/10.1007/s10845-024-02526-9.
5. B. Lane, I. Zhirnov, S. Mekhontsev, et al. Transient laser energy absorption, co-axial melt pool monitoring, and relationship to melt pool morphology. Additive Manufacturing 36 (2020) 101504, https://doi.org/10.1016/j.addma.2020.101504.
6. L.P. Zheng, Q. Zhang, H.Z. Cao, W.J. Wu, H.L. Ma, X.P. Ding, J. Yang, X.M. Duan, S.Q. Fan, Melt pool boundary extraction and its width prediction from infrared images in selective laser melting, Materials and Design. 183 (2019), 108110, https://doi.org/10.1016/j.matdes.2019.108110.
7. N.S. Selyutina, I.V. Smirnov, Yu.V. Petrov. Stabilisation effect of strain hysteresis loop for steel 45. International Journal of Fatigue, 2021, 145: 106133.
8. Y.F. Liu, S.Y Sun, J.L. Wang, et al. Tribological behaviors of LDED Inconel 718 samples polished with a hybrid laser polishing technique, Journal of Materials Research and Technology, 2023, 25: 633-646.

9. A. Heralica, A. Christianssona, B. Lennartson. Increased stability in laser metal wire deposition through feedback from optical measurements. Optics and Lasers in Engineering, 2010, 48: 478-485.
10. T. H. Kim, G. Y. Baek, J. B. Jeon, et al. Effect of laser rescanning on microstructure and mechanical properties of direct energy deposited AISI 316L stainless steel. Surface and Coatings Technology, 2021, 405: 126540.
11. R. Ma, C. Peng, Z. Cai, R. Wang, Z. Zhou, X. Li, X. Cao, Manipulating the microstructure and tensile properties of selective laser melted Al–Mg-Sc-Zr alloy through heat treatment, Journal of Alloys Compounds 831 (2020) 154773, https://doi.org/10.1016/j.jallcom.2020.154773.
12. Z. Wang, X. Lin, N. Kang, Y. Wang, X. Yu, H. Tan, H. Yang, W. Huang, Making selective-laser-melted high-strength Al–Mg–Sc–Zr alloy tough via ultrafine and heterogeneous microstructure, Scripta Materialia 203 (2021) 114052, https://doi.org/10.1016/j.scriptamat.2021.114052.
13. S. Ma, M. Jiang, X. Chen, B. Li, N. Jiang, Y. Chen, S. Wu, J. Liang, B. Li, Z. Lei, Y. Chen. Macro/micro-structure and mechanical properties of Al-6Mg-0.3Sc alloy fabricated by oscillating laser-arc hybrid additive manufacturing. Journal of Alloys Compounds. 929 (2022) 167325, https://doi.org/10.1016/j.jallcom.2022.167325.
14. K. Hao, Y. Gao, L. Xu, Y. Han, L. Zhao, W. Ren, Plasticity improvement coupled by carbon nanotubes and beam oscillation in laser-arc hybrid welding of magnesium alloy, Materials Science and Engineering A Structure 857 (2022) 144093, https://doi.org/10.1016/j.msea.2022.144093.
15. Clijsters S, Craeghs T, Buls S, Kempen K, Kruth JP. In situ quality control of the selective laser melting process using a high-speed, real-time melt pool monitoring system. Internation Journal of Advanced Manufaturing Technology 2014;75:1089–101. https://doi.org/10.1007/s00170-014-6214-8.
16. Y. Cai, Y. Wang, H. Chen, J. Xiong, Molten pool behaviors and forming characteristics in wire-laser directed energy deposition with beam oscillation, J. Mater. Process. Technol. 326 (2024) 118326, https://doi.org/10.1016/j.jmatprotec.2024.118326.
17. Wu Y, Fang J, Wu C, Li C, Sun G, Li Q. Additively manufactured materials and structures: a state-of-the-art review on their mechanical characteristics and energy absorption. Int J Mech Sci 2023;246:108102. https://doi.org/10.1016/j.ijmecsci.2023.108102.
18. Shevchik SA, Masinelli G, Kenel C, Leinenbach C, Wasmer K. Deep learning for in situ and real-time quality monitoring in additive manufacturing using acoustic emission. IEEE Trans Ind Inform 2019;15:5194–203. https://doi.org/10.1109/TII.2019.2910524.
19. Kanko JA, Sibley AP, Fraser JM. In situ morphology-based defect detection of selective laser melting through inline coherent imaging. J Mater Process Technol 2016;231:488–500. https://doi.org/10.1016/j.jmatprotec.2015.12.024.
20. Bartlett JL, Heim FM, Murty YV, Li X. In situ defect detection in selective laser melting via full-field infrared thermography. Additive Manufacturing 2018;24: 595–605. https://doi.org/10.1016/j.addma.2018.10.045.
21. C. Zhao, K. Fezzaa, R.W. Cunningham, H. Wen, F.D. Carlo, L. Chen, A.D. Rollett, T. Sun, Real-time monitoring of laser powder bed fusion process using high-speed X ray imaging and diffraction, Sci. Rep. 7 (2017) 3602, https://doi.org/10.1038/s41598-017-03761-2.
22. D. Ye, J.Y. Hsi Fuh, Y. Zhang, G.S. Hong, K. Zhu, In situ monitoring of selective laser melting using plume and spatter signatures by deep belief networks, ISA Trans. (2018), https://doi.org/10.1016/j.isatra.2018.07.021.
23. J.C. Heigel, B.M. Lane, Measurement of the melt pool length during single scan tracks in a commercial laser powder bed fusion process, J. Manuf. Sci. Eng. 140 (2018), https://doi.org/10.1115/1.4037571.
24. Y.Y. Su, Z.F. Wang, H.F. Lu, K.Y. Luo, J.Z. Lu, Improved wear resistance of directed energy deposited Fe-Ni-Cr alloy via closed-loop controlling laser power, J. Manuf. Process. 75 (2022) 802-813, https://doi.org/10.1016/j.jmapro.2022.01.047.

25. Bernauer C, Leitner P, Zapata A, Garkusha P, Grabmann S, Schmoeller M and Zaeh M F. 2024. Segmentation-based closed-loop layer height control for enhancing stability and dimensional accuracy in wire-based laser metal deposition. Robot. Comput. Int. Manuf. 86, 102683.
26. Zhao X, Lv Y, Dong S, Yan S, He P, Liu X, et al. The effect of thermal cycling on direct laser-deposited gradient H13 tool steel: microstructure evolution, nanoprecipitation behaviour, and mechanical properties. Mater Today Commun 2020;25. https://doi.org/10.1016/j.mtcomm.2020.101390.
27. N. Schnell, M. Schoeler, G. Witt, S. Kleszczynski, Experimental and numerical thermal analysis of the laser powder bed fusion process using in situ temperature measurements of geometric primitives, Mater. Des. 209 (2021), 109946.
28. K. Okten, Ä Biyikoglú, Development of thermal model for the determination of SLM process parameters, Opt. Laser Technol. 137 (2021), 106825.
29. Xiong J, Chen H, Zheng S M and Zhang G J. 2022. Feedback control of variable width in gas metal arc-based additive manufacturing. J. Manuf. Process. 76, 11–20.
30. Xiong J, Yin Z Q and Zhang W H. 2016. Closed-loop control of variable layer width for thin-walled parts in wire and arc additive manufacturing. J. Mater. Process. Technol. 233, 100–106.
31. Shi T, Lu B H, Shen T, Zhang R, Shi S H and Fu G Y. 2018. Closed-loop control of variable width deposition in laser metal deposition. Int. J. Adv. Manuf. Technol. 97, 4167–4178.
32. Mattera G, Caggiano A and Nele L. 2025. Optimal data driven control of manufacturing processes using reinforcement learning: an application to wire arc additive manufacturing. J. Intell. Manuf. 36, 1291–1310.
33. Anderson R. Characterization of direct metal deposition (DMD) bimetallic interfaces. The University of Alabama in Huntsville; 2018.
34. Li B, Jin Y, Yao J, Li Z, Zhang Q, Zhang X. Influence of laser irradiation on deposition characteristics of cold sprayed Stellite-6 coatings. Opt Laser Technol 2018;100:27–39.

Chapter 3
Effect of the Melt Pool In-Situ Control on Microstructure of Laser Directed Deposition 316L Stainless Steel Layers

3.1 Introduction

Directed energy deposition (DED) is a very promising additive manufacturing technique for the fabrication of sizable key components. DED offers a wide range of advantages namely a high level of formability in the manufacture of complex features, which can meet the demands of large-scale parts and be suitable for rapid manufacturing [1–3]. Consequently, DED has witnessed a rapid growth in many industries including automotive, aerospace, biomedical, and energy [4–6]. The metal powder is melted by high-energy laser beam to the surface of the substrate to form a melt pool with high temperature, high brightness, small size and fast change rate. Everton et al. [7] have proved that the formation of the melt pool helped to determine the grain growth in the solidification microstructure and was one of the main factors determining the surface quality of the coating, so the monitoring of the melt pool state is crucial [8–10]. The monitoring of the temperature distribution and geometry of the melt pool has become a hot topic in recent years. A process-mapping controlling melt pool size in laser-powder was developed by Vasinonta et al. [11], and the impact of building aspect ratio was explored. He et al. [12] recorded the geometrical characteristics of a Ti6Al4V melt cell based on the parameters of laser power and scanning speed. Yeung et al. [13] proposed the interpreter which utilized sophisticated LPBF laser control commands to improve dimensional accuracy and consistency of part quality.

Melt pool width (MPW) is thought as an important factor in melt pool shape, but its size acquisition mostly relies on simulation prediction, machine learning and offline monitoring. Scime et al. [14] measured multiple melt pool cross-sections of Inconel 718 alloy in the LPBF process for multiple combinations of process parameters. Cooka et al. [15] modelled the flow and heat transfer in and around the melt pool formed when melted the powder bed. Aggarwal et al. [16] used computational fluid dynamics of free surface thermal fluid simulations to predict the geometry of

K. Luo et al., *Melt Pool In-Situ Controlling of Laser Additive Manufacturing Technology*, https://doi.org/10.1007/978-981-95-4597-1_3

melt pools. Monitoring and controlling the melt pool width (MPW) could save manufacturing time and reduce overall cost by employing corrective actions to avoid undesirable results. So the on-line detection and real-time control of MPW is essential.

Thermal imaging is often considered to be an important feature to study thermophysical dynamics. In order to investigate the thermal behavior, Yang et al. [17] utilized the infrared camera test to get the temperature of the sample. A closed-loop control system was built based on the feedback of the infrared image sensing by Hu et al. [18]. Hofmeister et al. [19] used a high speed digital camera with a narrow bandpass optical filter to obtain thermal images of the LENS process zone.

The proportion integration differentiation (PID) optimization design method, considered as a feedback control algorithm, has been widely used in practical engineering applications. Wang et al. [20] investigated a model-based feed-forward control for laser power in LPBF to improve building quality. Li et al. [21] proposed that the control system using PID algorithm can improve part manufacturing quality based on the adaptive adjustment of process parameter. Zhong et al. [22] brought forward a layerwise feedback control strategy and model to ensure the quality of critical parts. Zheng et al. [23] used thermal imaging camera to monitor the surface temperature and predicted an average melt width.

In this chapter, the closed-loop control of MPW was proposed to ensure the cladding quality of complex components. Laser powers of 1600 W, 1800 W, and 2000 W were set as the initial powers. Two modes (traditional mode and control mode) were set as control groups to investigate the effects of closed-loop control of laser power on the melt pool width and microstructural evolution.

3.2 Experiment and Characterizations

3.2.1 LDED Process

The composition of the DED system was shown in Fig. 3.1a. The system consisted of CNC equipment, a fiber-coupled diode laser system, a 3D printing head, a Vacuum chamber and a gas-borne powder feeder. The coating material was Fe-Ni-Cr alloy (chemical composition wt%: Ni 12–13, Cr 16–18, Mo 2–3, Si 0.8, Mn 0.2, C 0.03, Fe Bal.) and the substrate was 45 steel. As seen in Fig. 3.1b, a coaxial infrared camera was mounted on the 3D print head to monitor the melt pool with real-time imaging. The device consists of a CLAMIR camera (high speed medium wavelength pbse- sensor camera) mounted coaxially on the laser head, an IR filter, (>700 nm), a set of optical mirrors, an iris, and an infrared notch-filter (1070 nm). The wavelength and resolution of the infrared camera are 1064 nm and 64 × 48 pixels, respectively.

In this study, straight single-track melt samples were prepared by a fixed laser scanning direction to study the melt pool morphology of the material, and

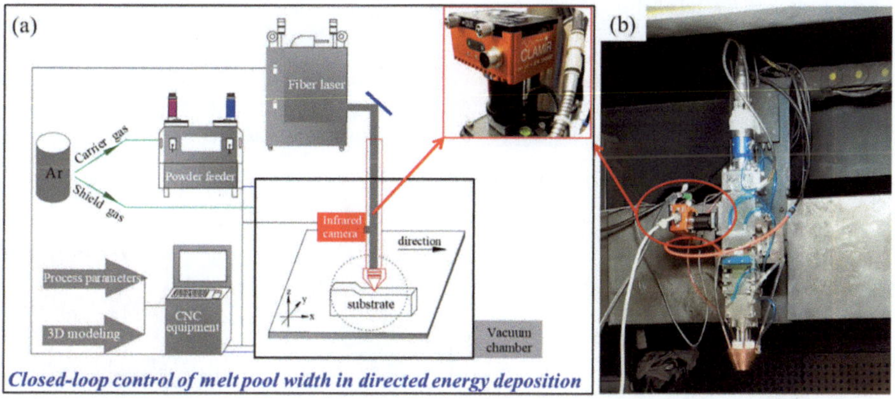

Fig. 3.1 The DED system with the closed-loop control of MPW. (**a**) Schematic diagram, and (**b**) physical picture

Table 3.1 The processing parameters in DED

Parameters	Value	Units
Powder feed rate	18	g/min
Scanning speed	10	mm/s
Overlapping rate	50%	/
Spot diameter	3	mm

single-layer melt samples were prepared by S-shaped laser scanning direction to study the bond strength and mechanical properties of the material. Laser powers ranging from 1000 W to 2400 W were set up for preliminary exploratory tests, with 200 W as a difference. Jiang et al. [24] found that the melt pool dilution rate was extremely low and the interfacial bonding was poor when the laser powers were 1000–1400 W; while Tian et al. [25] found that the large melt pool width and poor deposition quality were obtained when the laser powers were 2200 W–2400 W. Therefore, in order to ensure the forming quality, the laser powers of 1600–2000 W were selected as the control groups in this paper. The deposition parameters were shown in Table 3.1.

3.2.2 Melt Pool Control Strategy

Closed-loop controlled MPW method of the Fe-Ni-Cr alloy manufactured by DED was displayed in Fig. 3.2a. In an ideal thermal environment, the optimum MPW corresponded to the optimal part quality. During the DED process, the laser power preset in the traditional mode was switched on by the high energy laser to produce a complete melt pool on the substrate, which led to the corresponding MPW as the set-point in the control system. In the control mode, the difference between the

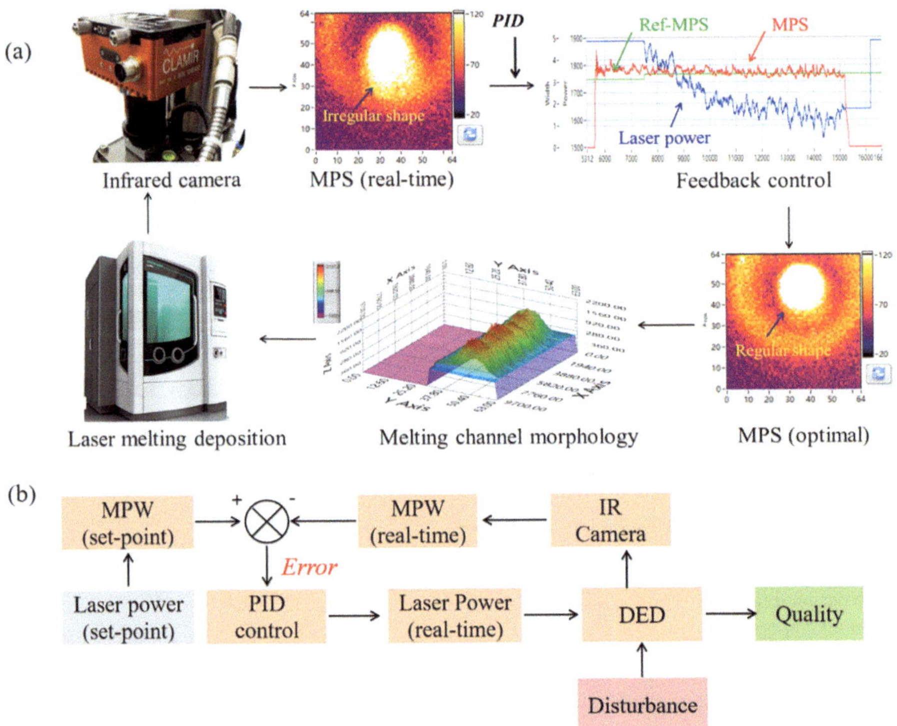

Fig. 3.2 Closed-loop controlled MPW method of Fe-Ni-Cr alloy. (**a**) Visual representation, and (**b**) block diagram

real-time MPW detected by the infrared camera and the set-point of the MPW was set as an input to the feedback control system. Based on the difference of the MPW, the laser power was adjusted by the PID algorithm to ensure the stability of the MPW and improve the quality of the parts. The response speed of the close-loop control system reached millisecond. The image acquisition and its processing rate was 1000 fps during the building process. So the whole process took 1 ms. In addition, during the closed-loop control process, the image acquisition, image processing and sending of analogue signals resulted in a time lag of ~50 ms, so a feed forward compensation was crucial.

In the closed loop control process, the melt pool image (Fig. 3.3a) taken by the infrared camera was converted into a MPW value by image processing as shown in Fig. 3.3. Luo et al. [26] has demonstrated that the gray-scale processing was the first step in analyzing the infrared image and obtaining the MPW (Fig. 3.3b) which prepared for the subsequent image processing. In order to avoid generating more interference information and ensure the better effect, Lu et al. [27] raised the importance of denoising. The mean filtering method was selected for de-manipulation processing (Fig. 3.3c), and an appropriate region of interest (ROI) was selected according to the current frame through the threshold segmentation method (Fig. 3.3d). The ROI contour was smoothed by the open operation in the morphological processing

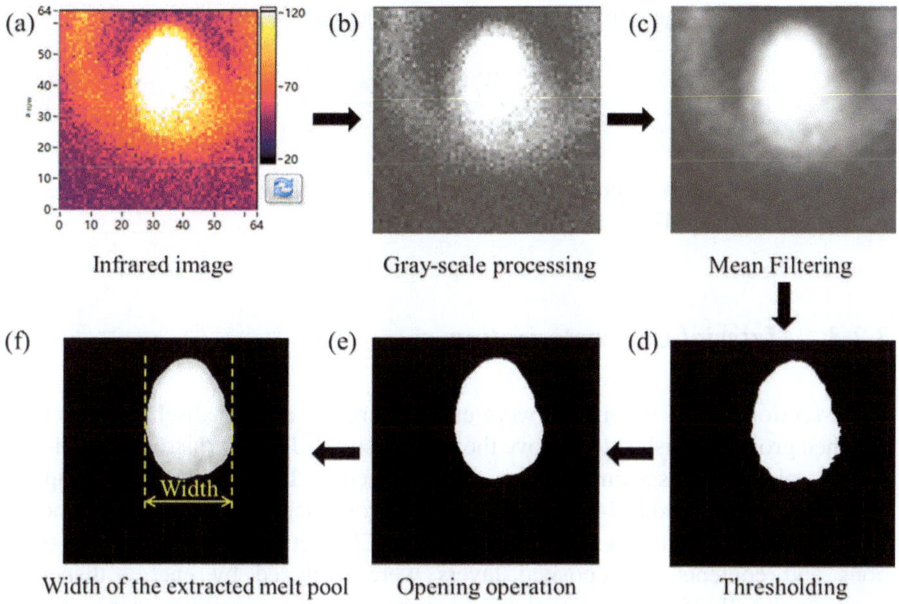

Fig. 3.3 The process for obtaining the MPW based on image processing methods (**a**) Infrared image, (**b**) gray-scale processing, (**c**) mean filtering, (**d**) thresholding, (**e**) opening operation, (**f**) image of melt pool width

(Fig. 3.3e), and finally the image of the width of MPW was obtained (Fig. 3.3f). The image resolution was set to 0.15 mm/pixel after the optical focusing was completed. Hooper [28] proposed to combine the software-calculated pixel values and working distances to obtain the MPW.

As shown in Fig. 3.2b, the closed-loop controlled MPW by PID algorithm included the following steps. Firstly, the difference between two widths was defined as

$$e(k) = W_i(k) - W_r(k) \tag{3.1}$$

where $W_i(k)$ was the first detected MPW as set-point value in the traditional mode, $W_r(k)$ was the MPW in real-time.

Secondly, the value of the single laser power variation was set to

$$\Delta P = K_P \left\{ e(k) + \frac{1}{T_I} \sum_{k}^{m=1} e(m) + T_D \left[e(k) - e(k-1) \right] \right\} \tag{3.2}$$

where the T_I was the integral time, K_p was the proportional gain, T_D was the differential time. The PID parameters were set as follows: K_P was 200, T_I was 0.5 and T_D was 0.5.

Thirdly, the change rate of the real-time detected MPW was defined as

$$\sigma = \frac{W_r(k) - W_i(k)}{W_i(k)} \tag{3.3}$$

When the melt width variation was controlled at ±3%, it was considered to be met the requirement of MPW stability.

3.2.3 *Material Characterization*

Cross-sections of all specimens were cut by wire-cut electric discharge machine, and then ground, polished to remove the surface stress. The microstructure and fracture morphology of specimens were characterized by LEICA MC170 HD optical microscopy (OM) and S-3400 N scanning electron microscopy (SEM). The porosity rate was measured with ImageJ software in SEM picture. The elemental distributions and contents of deposited layers were analyzed by energy dispersive spectrometer (EDS), and the phase composition was identified using the D8 ADVANCE X-ray diffractometer (XRD). The extracted samples for EBSD observation were shake-polished in an alumina suspension for 8 h after the standard metallographic polishing process. A Gemini SEM 500 was chose for electron backscattered diffraction (EBSD) analysis. Step size was kept constant at 0.2 μm, the operating voltage was 15 kV, and the acquired data were analyzed using Channel 5 software.

3.3 Microstructure Characterization for Single Track

3.3.1 *Melt Pool Width Control Results*

Figure 3.4 represents the changes of the width of melt pool and laser power based on 1600 W in two modes. It shows a single layer specimen, which is deposited in an "S" scan by 11 single tracks with a 50% overlapping rate. The MPS and laser power command which show the step changes are presented as well. The MPS and laser power are commanded at around 0 and approximately 60 cm. Laser power is modulated to achieve the desired MPS. Scanning speed does not decrease as the laser power decreases in this test, and beads are deposited at a constant deposition rate. The process remains within a stable range. The unit of scanning speed is set as mm/s during deposition, while the scanning path unit for the X-axis in Fig. 3.4 is set in cm. So there are many data points stacked on top of each other. When the initial laser power is 1600 W, as shown in Fig. 3.4, there is an obvious decrease tendency of laser power and the MPS with the rising scanning length after controller turning, while the MPS increases in the traditional mode without control. Specifically, the

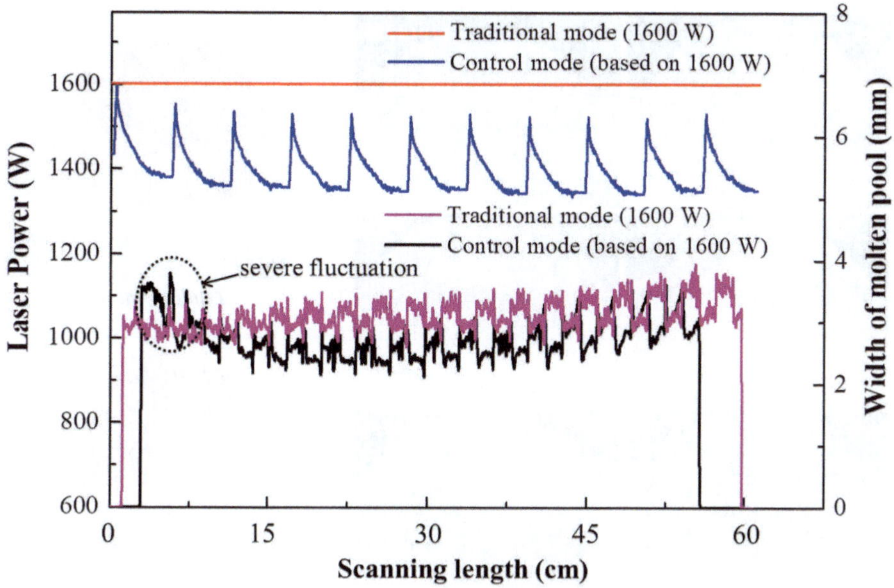

Fig. 3.4 Changes of the width of melt pool and laser power in two modes based on 1600 W

width of MPS gradually increases from ~2.9 to ~3.5 mm. However, when the controller turns on, the width of MPS gradually stabilized at 3 mm.

As presented in the black circle marked in the figure, it is noticeable that the severe fluctuation is observed early in the control mode, which is related to the synergy of three control components in Eq. (3.2). In the PID control process, the gain terms K_P, K_I and K_D are used to correct the laser power, correct the system error and slow down the oscillation of the melt pool, respectively. In the early stage of control process, the proportional gain K_P dominates, which accelerates the system response speed and obviously suppresses the melt pool width. In the case of the "S" scan mode and the 50% overlapping rate, the laser power is suddenly increased during the next deposition to compensate for the thermal effect. In terms of magnitude, laser power decreases from 1600 W to 1400 W, highlighting that the significant degree of changed laser power can influence MPS.

3.3.2 SEM and EBSD Observation

The cross sections of single tracks are shown in Fig. 3.5. The mechanism diagrams of dilution in two modes are respectively illustrated in Fig. 3.5g, h, and in accordance with the specified parameters in the figure proposed by Zhai et al. [29], the dilution is:

$$D = \frac{h_2}{h_1 + h_2} \times 100\% \tag{3.4}$$

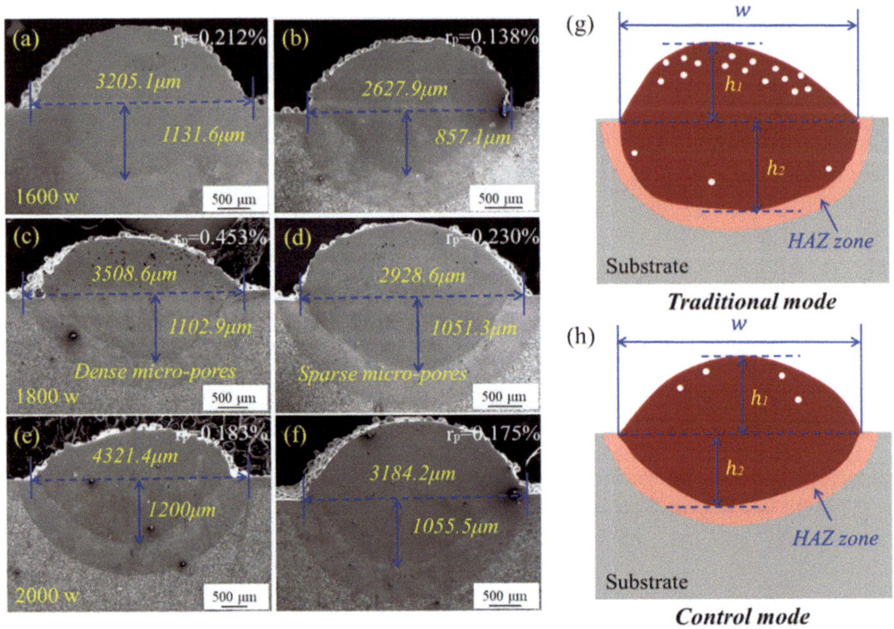

Fig. 3.5 Porosity statistics and cross-sectional view of the single-track melt pool morphology in both modes. (**a, c** and **e**) Dimension morphologies of melt pools at the traditional mode, (**b, d** and **f**) dimension morphologies of melt pools at the control mode, (**g** and **h**) mechanism diagrams of melt pool

where D is dilution of coating, h_1 is the height of the cladding layer, and h_2 is the melt pool depth in the substrate.

At 1600 W, the maximum depth of the melt pool in the substrate is 1131.6 μm and the MPW is 3205.1 μm in Fig. 3.5a. Moreover, the maximum depth and the width of the melt pool in Fig. 3.5b after control reach 857.1 μm and 2627.9 μm, respectively. According to Eq. (3.4), the dilution rates in two modes are 49.6% and 48.8%. Likewise, based on the 1800 W and 2000 W, the MPW and the h_2 are obviously smaller in the control mode than that in the traditional mode.

The dilution rates of specimens fabricated at 1800 W and 2000 W are 54.0% and 57.6% in the traditional mode (Fig. 3.5c, e), while the dilution rates in the control mode are 49.4% and 49.7% (Fig. 3.5d, f). The increase of the laser power will intensify the dilution of the melt pool, and closed-loop control method could effectively reduce the dilution rates. In the case of the high degree of bonding with the substrate, the low dilution rate is benefit to the improvement of the tensile property.

In addition to being related to the cooling rates and thermal gradients, the quality of the part formed is also related to the melt pool morphology, specifically manifests into physical defects in the three-dimensional build. The porosity rate can be written as $r_p = \frac{S_D}{S}$, where S_D is the total pixels of the porosity regions and S is the total pixels of the whole cladded layer. The porosities of Fig. 3.5a–f are 0.212%, 0.138%, 0.453%, 0.230%, 0.183%, 0.175%, respectively. As shown in Fig. 3.5g, in the traditional mode, most of the pores are densely distributed in the top of the cladding layer while the others distributed in the HAZ. Due to the increased melt pool temperature, the rapid vaporization of the liquid metal leads to the depression of the liquid surface and the collapse caused by water vapor drops. Bubbles and organic matter have no time to escape, making it easy to form porous defects at the top of the deposition layer. In the traditional mode, the increasing melt width enhances the Rayleigh-plateau instability, which is another reason for the formation of pores. As shown in Fig. 3.5h, the number of pores is greatly reduced in the control mode. Because the reduced laser power slows down the rate of liquid metal vaporization and water vapor collapse, further reducing porosity defects. Moreover, Yadroitsev et al. [30] reported that the Rayleigh-plateau instability was also reduced by smaller melt width, thus reducing the occurrence of defects.

The EBSD analysis in the central region of the specimen for two modes is presented in Fig. 3.6. In the traditional mode (Fig. 3.6a, c), the microstructure is mainly characterized by large dendrites with consistent growth direction. The grain size of specimen is close to normal distribution and has an average value of 1.90 μm. However, the microstructure in the control mode (Fig. 3.6b) is significantly different from that. It is consisted of fish scaly structure, and the grain size of specimen has an average value of 1.58 μm which is typical equiaxed crystals. Comparing the phases distribution in Fig. 3.6e with that in Fig. 3.6f, it is clearly that austenite (γFe, Ni) phase is nearly spherical dispersedly located at grain boundary, and the content of γ phase in the control mode is obviously increased. It is proved that the grain refinement could be promoted by increasing austenite content.

When the laser power is 1600 W, cross-sectional views of single tracks in the DED process are shown in Fig. 3.7a, b, which clearly present the contours of the longitudinal melt pool boundary. In the traditional mode, the uneven melt pools

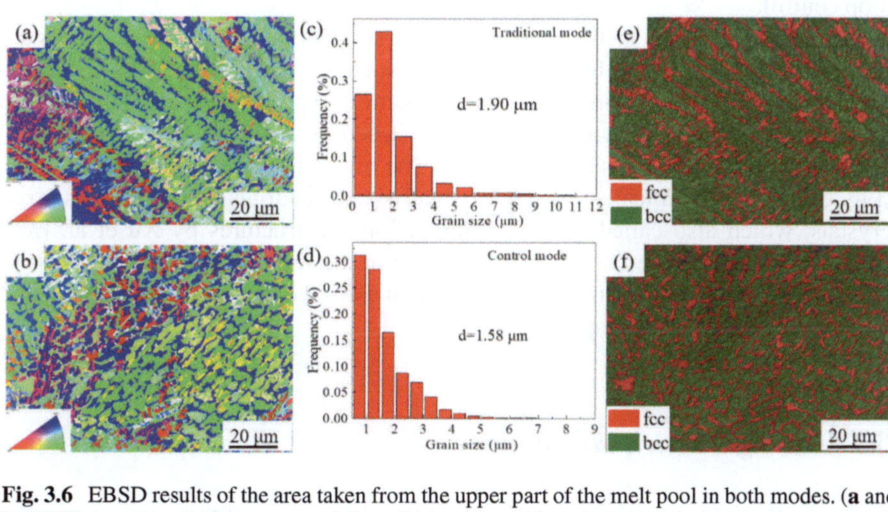

Fig. 3.6 EBSD results of the area taken from the upper part of the melt pool in both modes. (**a** and **b**) EBSD inverse pole figure maps of depositions, (**c** and **d**) statistical analysis of grain size distributions, (**e** and **f**) fcc and bcc phases distributions

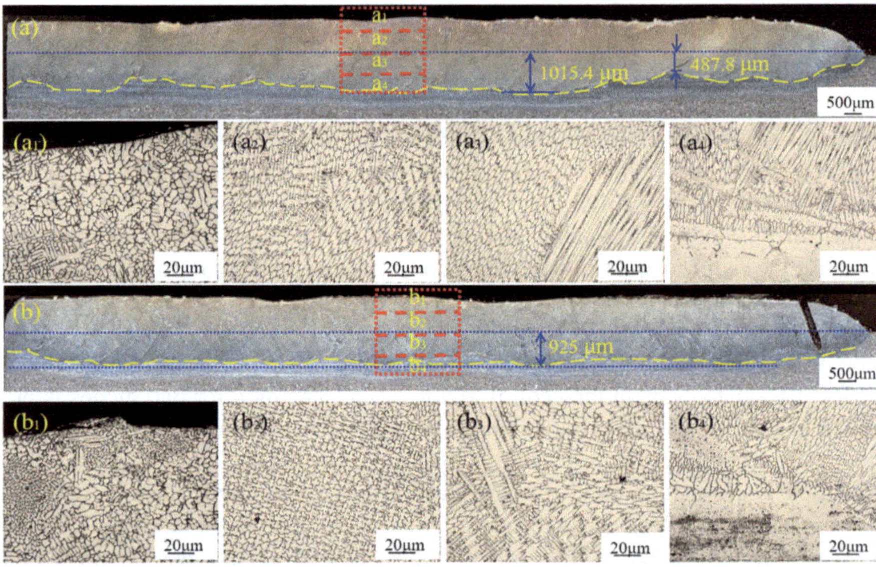

Fig. 3.7 Cross-sectional views of single tracks along the deposited direction in two modes. (**a**) The traditional mode, (**b**) the control mode, (**a_1–a_4**) OM observation of microstructure from top to bottom of single track at high magnification, (**b_1–b_4**) OM observation of microstructure from top to bottom of single track at high magnification

exist in the single track, reaching the maximum depth of ~1015.4 μm and the minimum depth of ~487 μm, with a depth variation of ~52%. Uniform depth at the bottom of the melt pool with a maximum depth of ~925 μm is achieved in the control mode. Clearly, the longitudinal melt pool is flatter and more uniform under closed-loop control.

The microstructures of the areas deposited at different depths in two modes are shown in Fig. 3.7a_1–a_4, b_1–b_4. Results show that closed-loop controlled MPW has a pronounced effect on the microstructural evolution, especially in the central region of the single track. As shown in Fig. 3.7a_2, a_3, structures are composed of large-sized dendrites and typical fish scaly structures. The microstructure in the control mode (Fig. 3.7b_2, b_3) specifically exhibit as small-sized dendrites and dense equiaxed crystals, which also called typical solidification substructures by Xu et al. [31]. There is little influence on the microstructure at the top and bottom of the single track after closed-loop controlled MPW, which exhibit dense equiaxed crystals and growing planar crystals, respectively.

3.3.3 Single Track Morphology

The overall surface morphologies of single tracks during DED in two modes are presented in Fig. 3.8. In the traditional mode (Fig. 3.8a, c, e), the morphologies of the deposited surface based on three laser powers show the rough and non-uniform

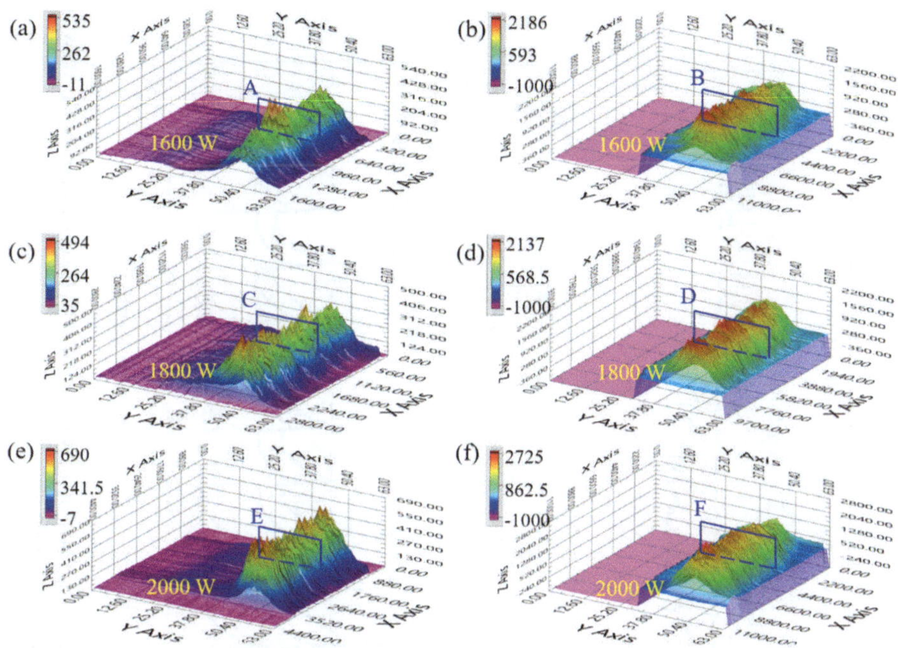

Fig. 3.8 Single track morphologies based on three laser powers. (**a**, **c** and **e**) The traditional mode, (**b**, **d** and **f**) the control mode

state as the scanning length increases. In addition, there is a tendency for the bead height to increase gradually with increasing laser power. However, after closed-loop control, as shown in Fig. 3.8b, d, f, the beads are relatively uniform in width and consistent in height. The melt pool width is effectively suppressed, and the uniformity and flatness of the surface profile of the monorail is significantly improved.

3.4 Microstructure Characterization for Single Layer

3.4.1 Melt Pool Width Control Results

Figure 3.9 shows the step change in laser power and MPW for deposited single layer samples in both modes. It can be found that the point at which the laser turns on and turns off are marked in Fig. 3.9a–c. In order to determine whether the control requirements are met in the control mode, blue dashed lines are set up in Fig. 3.9a–c to indicate the ideal melt pool width. At 1600 W (Fig. 3.9a), the preset MPW of the deposition process is 3 mm. With the increase of scanning length, the MPW in the traditional mode gradually increases from 3 to 3.5 mm. In the control mode, the MPW decreases obviously and finally reaches the 3 mm, indicating the effective control of MPW. Similarly, the MPW increases continuously at laser powers of 1800 W (Fig. 3.9b) and 2000 W (Fig. 3.9c) in the traditional mode. The original

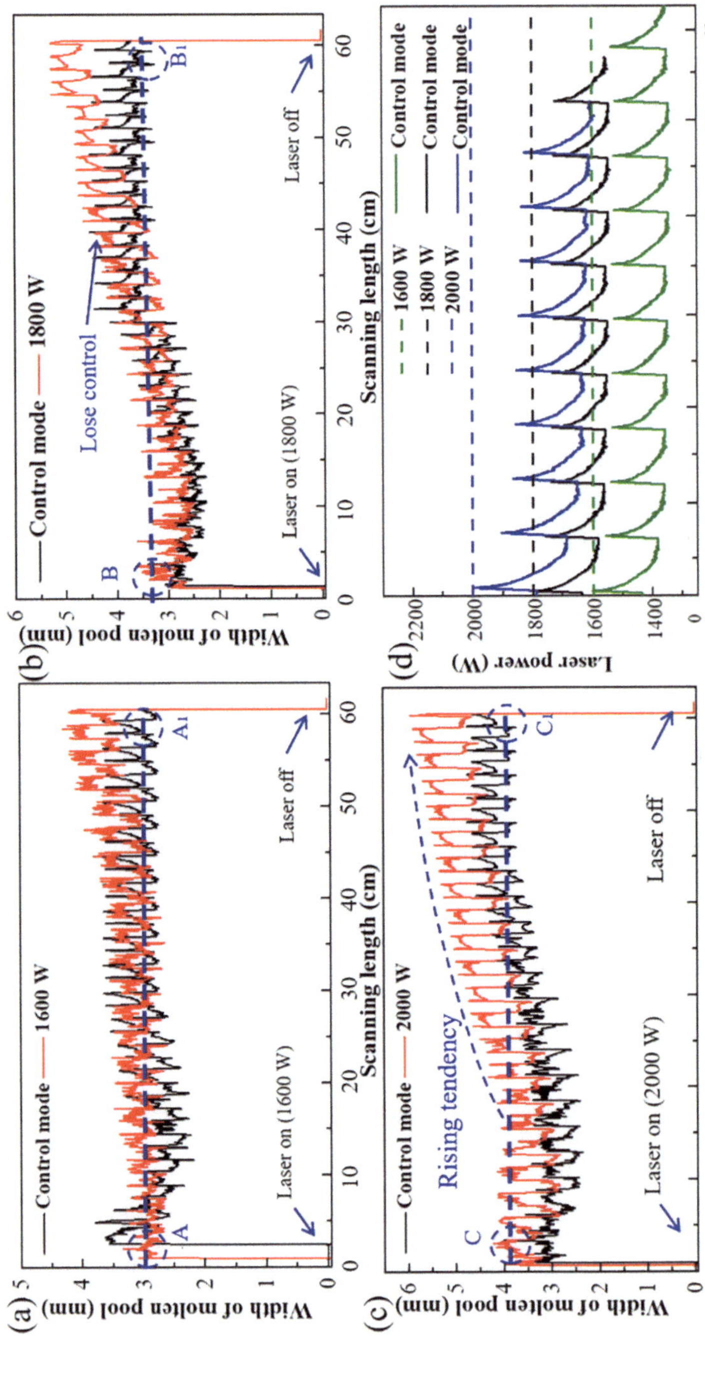

Fig. 3.9 Step changes in laser power and MPW of three laser powers in two modes. **(a)** Comparison of the MPW at 1600 W, **(b)** comparison of the MPW at 1800 W, **(c)** comparison of the MPW at 2000 W, **(d)** change of three laser powers in the control mode

MPW detected are 3.4 mm and 4 mm, respectively. As can be seen from points B_1 and C_1, under the closed-loop control, the changes of MPW are kept within ±3% variation of the optimal ones at the later stages of deposition, which indicates the desired control target has been achieved. In Fig. 3.9d, the dashed line represents the constant power in the traditional mode and the solid line represents the dynamically varying power in the control mode. The laser power tends to decrease slowly under the closed loop control of the PID algorithm. The laser power decreases from 1600 W to ~1344 W, 1800 W to ~1550 W and 2000 W to ~1590 W, respectively.

It can be found that the melt pool width eventually reaches 3.5 mm in both the 1600 W open-loop case and the 1800 W closed-loop case, but the trend of curves of laser power and melt pool width in these two cases are totally different. In the 1600 W open-loop case, the laser power remains constant and the melt pool width increases continuously from 3 mm to 3.5 mm. In the 1800 W closed-loop case, the laser power is gradually reduced from 1800 W to ~1550 W. Due to the stepwise reduction in laser power, the melt pool width reduces from 3 mm to ~2.2 mm at first and then gradually increases to ~3.5 mm. Finally, the melt pool width is stabilized at 3% of the set value.

Considering a control volume that encompasses the entirety of the liquid melt pool and heat affected zone, the total input laser energy, P_{in}, can be assumed to be either absorbed into the control volume (P_{abs}), or reflected off the surface (P_{ref}) as light energy at the laser wavelength. So the total laser input (P_{in}) can be defined as

$$P_{\mathrm{in}} = P_{abs} + P_{ref} \tag{3.5}$$

Lane et al. [32] considered all energy conversions that were not reflected by the laser energy source to be P_{abs}. In the traditional mode, except for the first detected MPW during deposition, each subsequent state could be influenced by the reflected laser light from the surface in the previous state. When the scanning speed and powder feed rate are constant, the melted metal surface increases due to the heat accumulation, leading to the gradually increased MPW.

In addition, it can be noticed that the setting of the MPW increases with increasing laser power, which is related to the flow state of the melt pool. Aggarwal et al. [16] considered that the effect of the impact of the powder particles on the hydrodynamic and thermal behavior of the melt pool, the resulting oscillations and instabilities were usually analyzed by the Peclet number (Pe). The Peclet number is the ratio of the convective to the conductive heat transfer which defined as

$$Pe = \frac{\rho_l U_{avg} C_{pl} L_R}{k_l} \tag{3.6}$$

where ρ_l is the density of molten Fe-Ni-Cr alloy, C_{pl} is the specific heat, U_{avg} is the melt pool average velocity, L_R is the half width of the melt pool and $k_l$is the molten metal thermal conductivity. The Peclet number is proportional to the laser power. As the laser power increases, in the case of increased Peclet number and molten metal thermal conductivity, and the case of the constant density of molten Fe-Ni-Cr alloy, melt pool average velocity and specific heat, the MPW increases. Furthermore, it is also related to the HAZ. Due to the larger HAZ, more feeding powder is melted with the increase of laser power.

It can be noticed that the MPW curves in the control mode are decrease and then slowly increase, which is related to the PID control in Eq. (3.2). In the early stage, the proportional gain K_p plays a key role, accelerating the system response and significantly suppressing MPW. In the later stages, the integral gain K_I and the differential gain K_D are strengthened, which can effectively reduce the suppression effect of the MPW and slow down the oscillation. In addition, the control stability could be improved until reaching the target width.

When deposited the single layer, the curves of the changed laser power based on 1600 W–2000 W are shown in Fig. 3.9d. Each repetition of laser power indicates the formation of a track. In the control mode, in order to suppress the infinite increase in MPW, the laser powers are reduced by the feedback control system. The sudden increase in laser power at the end of the deposited single layer is to compensate for thermal effects.

3.4.2 OM and SEM Observation

In both modes, the complete morphology and microstructure of the melt pool, perpendicular to the deposition direction, is shown in Fig. 3.10. As can be seen in Fig. 3.10a, b, the grain size has been further reduced after closed-loop control, which can be evidenced by the transformation of the large dendritic crystals running through the melt pool into a dense distribution of equiaxed crystals and small cell crystals. In two modes, the microstructures from the top to the bottom of the melt pool are shown in Fig. 3.10c–g, respectively. Bermingham et al. [33] reported that the large temperature gradient and low solidification rate resulted in the microstructure of planar crystals at the bottom.

In the traditional mode, due to the rapid heat accumulation, the cooling rate decreases and the solidification rate increases. The middle grains form dendrites with the same growth direction along the heat dissipation direction, which approximately perpendicular to the lap fusion line with the characteristics of epitaxial growth. Jian et al. [34] have investigated that the dendritic structures were slightly coarsened at every two lap heat-affected places. In addition, there are some cell crystals with austenite characteristics around the large-size dendrites. After control, the cooling rate and solidification rate of the melt pool are accelerated by reducing the laser power input. The structure is dominated by uniform equiaxed crystals with small size.

The XRD pattern analysis is illustrated in Fig. 3.11. From the XRD pattern for specimens in two modes, the main phases all belong to ferrite (αFe, Cr) and austenite (γFe, Ni). In the traditional mode, the ferrite (αFe, Cr) is the main phase, the generation of a certain amount of austenite (γFe, Ni) restrict the diffusion of ferrite

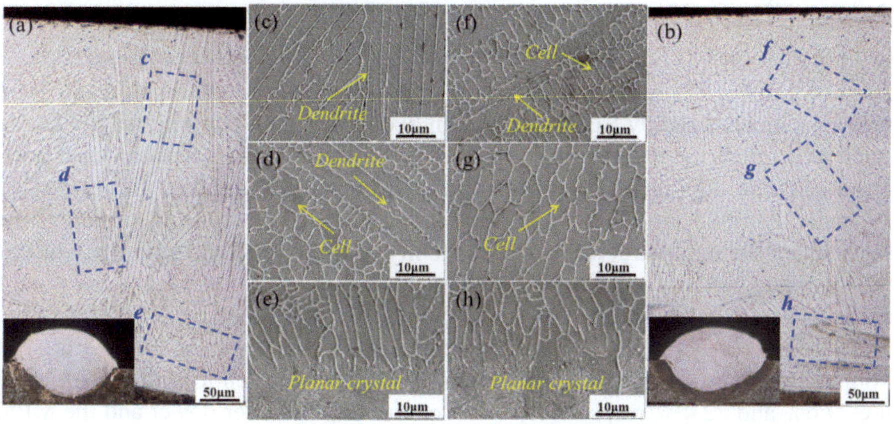

Fig. 3.10 OM and SEM observations of the cross-section profile and microstructure of melt pool in two modes. (**a**, **c**–**e**) the traditional mode, (**b**, **f**–**h**) the control mode

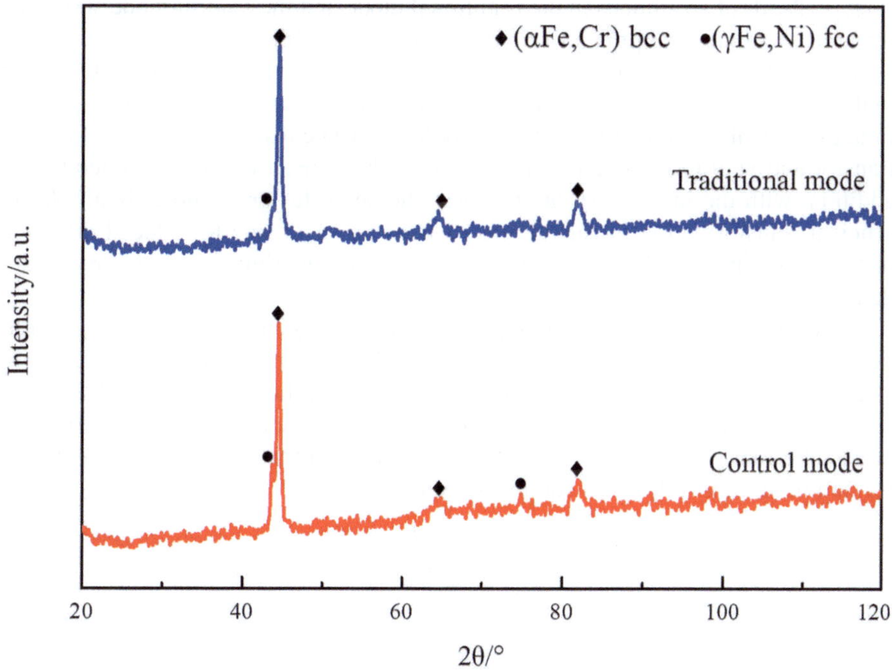

Fig. 3.11 XRD pattern of Fe-NiCr alloy in two modes

(αFe, Cr). In the control mode, the XRD pattern for alloy indicates the obvious increase of austenite (γFe, Ni). In addition, the stronger intensity of austenite leads to the distinct austenite solidification structure when forms in super cooling solidification.

3.4.3 EBSD Observation

As shown in Fig. 3.12, microstructure morphology, phase transformation and grain size are characterized by EBSD analysis in two modes. In the traditional mode, the microstructure is mainly manifested as a densely distributed lath ferrite while small-sized cells are dominant in the control mode. Figure 3.12c shows the ferrite and austenite distributions of Fe-Ni-Cr alloy in the traditional mode. According to the phase map, the microstructure consists mainly of α-phase ferrite with the lamellar structure and partially transformed γ-phase austenite, where the volume fraction of ferrite is 81.4% and austenite is 18.6% [35]. IPF maps obtained by EBSD for ferrite and austenite of Fe-Ni-Cr alloy in the control mode are shown in Fig. 3.12d. The volume fractions of ferrite with lamellar α structure and austenite with cell structure are 77.6% and 22.4%, respectively. After controlling the laser power and the MPS, the volume fraction of austenite increases significantly due to the improved transformation of ferrite to austenite, which also leads to the formation of more cell structures and grain refinement. As can be seen in Fig. 3.12e, f, the IPF maps (Fig. 3.12c, d) of specimens in the traditional mode and the control mode show that the grain sizes are 1.63 μm and 1.56 μm, respectively [36, 37].

The differences in microstructure appearing in the two modes are closely related to the phase transformation after controlled power. In the traditional mode, the continuous thermal accumulation and slow cooling rate at constant laser power limit the transformation of ferrite to austenite, and more dendrite structures are generated. In addition, with the increase of laser power, the dendrites grow more tightly. More efficient transformation of tighter spaced laths proceeds due to reduced diffusion distances. Ultimately, the residual ferrite cuts across the original growth direction of dendrites and cell crystals.

In the control mode, due to the separation of the alloying elements during solidification, the diffusion rate of these elements at high temperatures is relatively low. The closed-loop control of the laser power speeds up the cooling rate of the melt pool which causes the further increase of austenite content. It limits the growth of large slate-like grains during super cooling solidification and finally leads to the formation of uniformly sized equiaxed crystals.

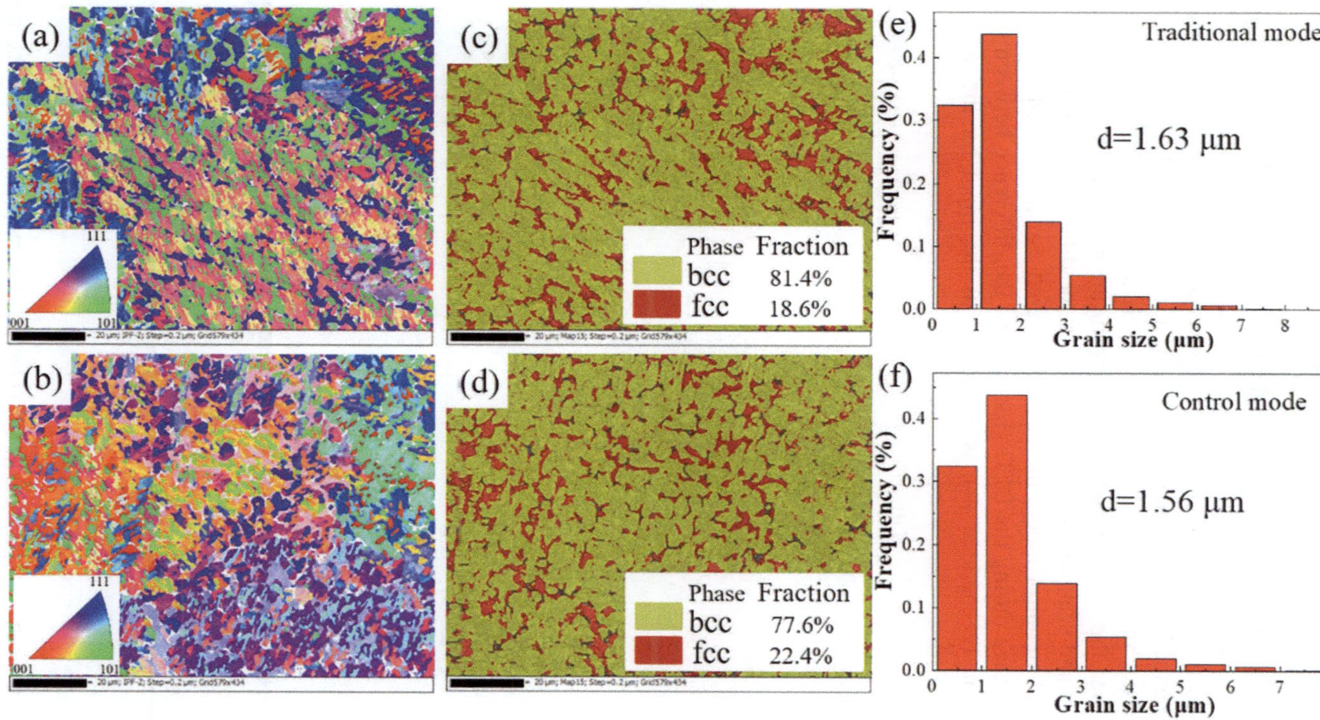

Fig. 3.12 EBSD analysis in the traditional mode (**a**, **c**, **e**) and the control mode (**b**, **d**, **f**). (**a**) and (**b**) EBSD inverse pole figure (IPF-Z) maps of depositions, (**c**) and (**d**) IPF maps for ferrite and austenite of the analyzed area, (**e**) and (**f**) statistical analysis of grain size distributions

3.5 Influence Mechanisms of Melt Pool Control on the Microstructure

Figure 3.13 represents the schematic diagrams of DED process, the microstructure mechanism and the longitudinal melt pool boundary. In the traditional mode, the powder is directly melted to the substrate using constant and high-energy laser and then forms the melt pool, as shown in Fig. 3.13a. During the DED, as the scanning length increases, the constant laser power accelerates the accumulation of heat, which increases the average melt pool depth in the substrate. When the powder feeding speed and scanning speed are constant, the dilution rate of the melt pool increases. In the control mode, the coaxial infrared camera is used to detect the molten pool image during the DED process, and the difference between the real-time MPW and the optimal one is calculated as the input for the subsequent

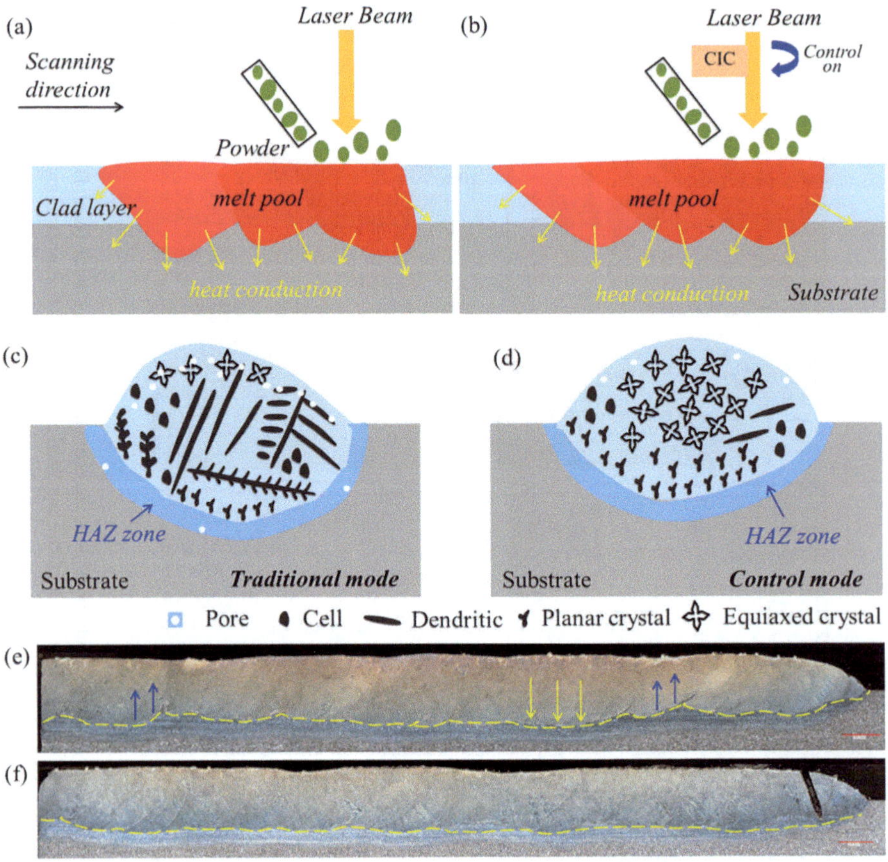

Fig. 3.13 Schematic diagrams of MPW in DED process in two modes. (**a**) and (**b**) The formation process of the melt pool, (**c**) and (**d**) microstructure and defect distribution of melt pool in two modes, (**e**) and (**f**) cross-sectional views of single track in two modes

adjustment of the laser power. After a large number of engineering tests, Su et al. [38] reported that appropriate PID parameters and the laser power can be adjusted to correct the MPW. After controlling the laser power, the decrease of heat input accelerates the cooling rate, and the smaller average melt pool depth in the substrate leads to the lower dilution rate. On the premise that the cladding layer and the substrate are well combined, the mechanical properties of the components are significantly improved.

In two modes, the microstructure and defect distribution of the melt pool are presented in Fig. 3.13c, d. Due to the accumulation of heat and the slow heat dissipation rate in the traditional mode, the middle crystal grains form large-size dendrites along the heat dissipation direction in the same growth direction. When analyzed in conjunction with the XRD results, these types of morphology are ferrite and austenite. Shen et al. [39] have confirmed that due to the restricted diffusion, the dendritic crystals formed in the super cooling solidification during the ferrite austenite transformation. In addition, Zhang et al. [40] proved that the bubbles were too late to overflow with the increasing of the temperature of the melt pool, and many densely distributed pores were formed in the top of the cladding layer and the HAZ. In the control mode, the less superimpose of the temperature of the melt pool leads to the fast cooling rate. The high nucleation rate makes the structure without segregation is easy to form. Combined with the analysis of XRD results, more austenite content is transformed from ferrite. Vecchiato et al. [41] noted that the increase in austenite content limited the growth of large dendrites, resulting in uniformly sized equiaxed crystals. In the control mode, the reduced laser power slows down the rate of liquid metal vaporization and water vapor collapse, and the Rayleigh-plateau instability also be reduced due to the smaller MPW, thus reducing the occurrence of defects.

In addition, the longitudinal melt pool boundaries in two modes are presented in Fig. 3.13e, f. Compared with the single track in the traditional mode, it is easily found that both the surface of the cladding layer and the average melt pool depth in the substrate are more uniform and flat in the control mode. Furthermore, the transient areas at the end of the track, which would reduce the depth, are obviously controlled under closed-loop control.

3.6 Summary

In the chapter, the closed-loop control method was proposed for the first time to precisely adjust the melt pool width in large-scale directed energy deposition of Fe-Ni-Cr alloy. It directly took the difference between the real-time MPW detected by the infrared camera and the initial MPW in the traditional mode as the input of the PID algorithm, and then adjusted the MPW by controlling the laser power to achieve the online control of part quality. Some important conclusions were listed as below:

1. Compared with that in the traditional mode, the dilution rate of the melt pool and pore defects significantly reduced in the control mode, which further improved the flatness of the coating and the uniformity of the single track.

2. The closed-loop control method suppressed the thermal build-up in the melt pool by regulating the laser power, thus promoting the production of more austenite. The increase in austenite content was the main reason for the development of more equiaxed crystals.
3. The closed-loop control method effectively regulated the heat accumulation and cooling rate of the melt pool, which also accelerated the transformation of ferrite to austenite. The monitoring system not only suppressed the growth of large dendrites, but also promoted the formation of equiaxed crystals.

References

1. T. Wang, Y.Y. Zhu, S.Q. Zhang, H.B. Tang, H.M. Wang, Grain morphology evolution behavior of titanium alloy components during directed energy deposition additive manufacturing, J. Alloys Compd. 632 (2015) 505–513, https://doi.org/10.1016/j.jallcom.2015.01.256.
2. T. DebRoy, H.L. Wei, J.S. Zuback, T. Mukherjee, J. W. Elmer, J. O. Milewski, A. M. Besee, A. W. Heid, A.De, W.Zhang, Additive manufacturing of metallic components - process, structure and properties, Prog. Mater. Sci. 92 (2018) 112-224, https://doi.org/10.1016/j.pmatsci.2017.10.001.
3. Q. Liu, Y.D. Wang, H. Zheng, K. Tang, L. Ding, H.X. Li, S.L. Gong, Microstructure and mechanical properties of LMD–SLM hybrid forming Ti6Al4V alloy, Mater. Sci. Eng. A.660 (2016) 24-33, https://doi.org/10.1016/j.msea.2016.02.069.
4. A. Emelogu, M. Marufuzzaman, S.M. Thompson, N. Shamsaei, L. Bian, Additive manufacturing of biomedical implants: a feasibility assessment via supply-chain cost analysis, Addit. Manuf. 11 (2016) 97–113, https://doi.org/10.1016/j.addma.2016.04.006.
5. L. Columbus, The state of 3D printing, Forbes 2017 (2017).
6. D.L. Bourell, M. Leu, D. Rosen, Roadmap for additive manufacturing-Identifying the future of freeform processing. The University of Texas at Austin, Laboratory for Freeform Fabrication, Adv. Manuf. Cent. (2009) 32.
7. S.K. Everton, M. Hirsch, P. Stravroulakis, R.K. Leach, A.T. Clare, Review of in-situ process monitoring and in-situ metrology for metal additive manufacturing, Mater. Des. 95 (2016) 431–445, https://doi.org/10.1016/j.matdes.2016.01.099.
8. T.G. Spears, S.A. Gold, In-process sensing in selective laser melting (SLM) additive manufacturing, Integr. Mater. Manuf. Innov. 5 (1) (2016) 16–40.
9. B.T. Gibson, Y.K. Bandari, B.S. Richardson, W.C. Henry, E.J. Vetland, T.W. Sundermann, L.J. Love, Melt pool size control through multiple closed-loop modalities in laser-wire directed energy deposition of Ti-6Al-4V, Addit. Manuf. 32 (2020) 100993, https://doi.org/10.1016/j.addma.2019.100993.
10. J. Haley, C. Leach, B. Jordan, R. Dehoff, V. Paquit, In-situ digital image correlation and thermal monitoring in directed energy deposition additive manufacturing, Optics. Express. 29 (2021) 9927-9941, https://doi.org/10.1364/OE.416659.
11. A. Vasinonta, J. Beuth, M. Griffith, Process maps for controlling residual stress and melt pool size in laser-based SFF processes, Proceedings of the Solid Freeform Fabrication Symposium (2000) 200–208.
12. Y.N. He, C. Montgomery, J. Beuth, B. Webler, Melt pool geometry and microstructure of Ti6Al4V with B additions processed by selective laser melting additive manufacturing, Mater. Des. 183 (2018)108216, https://doi.org/10.1016/j.matdes.2019.108126.
13. H. Yeung, B.M. Lane, M.A. Donmez, J.C. Fox, J. Neira, Implementation of advanced laser control strategies for powder bed fusion systems, Procedia Manuf. 26 (2018) 871–879, https://doi.org/10.1016/j.promfg.2018.07.112.

14. L. Scime, J. Beuth, Melt pool geometry and morphology variability for the Inconel 718 alloy in a laser powder bed fusion additive manufacturing process, Addit. Manuf. 29 (2019) 100830, https://doi.org/10.1016/j.addma.2019.100830.
15. P.S. Cooka, A.B. Murphy, Simulation of melt pool behaviour during additive manufacturing: Underlying physics and progress, Addit. Manuf. 31 (2020) 100909, https://doi.org/10.1016/j.addma.2019.100909.
16. A. Aggarwal, A. Chouhan, S. Patel, D.K. Yadav, A. Kumar, A.R. Vinod, K.G. Prashanth, N.P. Gurao, Role of impinging powder particles on melt pool hydrodynamics, thermal behaviour and microstructure in laser-assisted DED process: A particle-scale DEM – CFD – CA approach, Int. J. Heat Mass Transf. 158 (2020) 119989, https://doi.org/10.1016/j.ijheatmasstransfer.2020.119989.
17. D.Q. Yang, G. Wang, G.J. Zhang, Thermal analysis for single-pass multi-layer GMAW based additive manufacturing using infrared thermography, J. Mater. Process. Technol. 244 (2017) 215–224, https://doi.org/10.1016/j.jmatprotec.2017.01.024.
18. D. Hu, H. Mei, G. Tao, R. Kovacevic, Closed loop control of 3d laser cladding based on infrared sensing, Proceedings of the Solid Freeform Fabrication Symposium (2001) 129–137.
19. W.H. Hofmeister, D.O. MacCallum, G.A. Knorovsky, Video monitoring and control of the LENS process, Proceedings of the 9th International Conference of Computer Technology in Welding, Detroit, MI, 1999, pp. 28–30 Sept.
20. Q. Wang, P. Michaleris, A.R. Nassar, J.E. Irwin, Y. Ren, C.B. Stutzman, Model-based feed-forward control of laser powder bed fusion additive manufacturing, Addit. Manuf. 31 (2020) 100985, https://doi.org/10.1016/j.addma.2019.100985.
21. J.Y. Li, et al, An extended lumped-parameter model of melt-pool geometry to predict part height for directed energy deposition, J. Manuf. Sci. Eng. 139.9 (2017) 091016.
22. Q. Zhong, X.Y. Tian, X.K. Huang, C.B. Huo, D.C. Li, Using feedback control of thermal history to improve quality consistency of parts fabricated via large-scale powder bed fusion, Addit. Manuf. 42 (2021) 101986, https://doi.org/10.1016/j.addma.2021.101986.
23. L.P. Zheng, Q. Zhang, H.Z. Cao, W.J. Wu, H.L. Ma, X.P. Ding, J. Yang, X.M. Duan, S.Q. Fan, Melt pool boundary extraction and its width prediction from infrared images in selective laser melting, Mater. Des. 183 (2019) 108110, https://doi.org/10.1016/j.matdes.2019.108110.
24. C.Y. Jiang, Y. Xie, C. Kong, J.Q. Zhang, D.J. Young, Corrosion behaviour of Ni-Cr alloys in mixed oxidising gases at 650 °C, Corros. Sci. 174 (2020) 108801, https://doi.org/10.1016/j.corsci.2020.108801.
25. J.Y. Tian, P. Xu, Q.B. Liu, Effects of stress-induced solid phase transformations on residual stress in laser cladding a Fe-Mn-Si-Cr-Ni alloy coating, Mater. Des. 193 (2020) 108824, https://doi.org/10.1016/j.matdes.2020.108824.
26. C. Luo, et al., Finite element analysis of temperature and stress fields during the selective laser melting process of thermoelectric SnTe, J. Mater. Process. Technol. 261 (2018) 74e85, https://doi.org/10.1016/j.jmatprotec.2018.06.001.
27. Q.Y. Lu, N.V. Nguyen, A.J.W. Hum, T. Tran, C.H. Wong, Optical in-situ monitoring and correlation of density and mechanical properties of stainless steel parts produced by selective laser melting process based on varied energy density, J. Mater. Process. Technol. 271 (2019) 520–531, https://doi.org/10.1016/j.jmatprotec.2019.04.026.
28. P.A. Hooper, Melt pool temperature and cooling rates in laser powder bed fusion, Addit. Manuf. 22 (2018) 548-559, https://doi.org/10.1016/j.addma.2018.05.032.
29. L.L. Zhai, C.Y. Ban, J.W. Zhang, Microstructure, microhardness and corrosion resistance of NiCrBSi coatings under electromagnetic feld auxiliary laser cladding, Surf. Coat. Technol. 358 (2019) 531–538, https://doi.org/10.1016/j.surfcoat.2018.11.034.
30. I. Yadroitsev, A. Gusarov, I. Yadroitsava, I. Smurov, Single track formation in selective laser melting of metal powders, J. Mater. Process. Technol. 210 (2010) 1624-1631, https://doi.org/10.1016/j.jmatprotec.2010.05.010.
31. X. Xu, H.F. Lu, K.Y. Luo, J.H. Yao, L.Z. Xu, J.Z. Lu, Y.F. Lu, Mechanical properties and electrochemical corrosion resistance of multilayer laser cladded Fe-based composite coatings on

4Cr5MoSiV1 steel, J. Mater. Process. Technol. 284 (2020) 116736, https://doi.org/10.1016/j.jmatprotec.2020.116736.
32. B. Lane, I. Zhirnov, S. Mekhontsev, S. Grantham, R. Ricker, S. Rauniyar, K. Chou, Transient laser energy absorption, co-axial melt pool monitoring, and relationship to melt pool morphology, Addit. Manuf. 36 (2020) 101504, https://doi.org/10.1016/j.addma.2020.101504.
33. M.J. Bermingham, D.H. StJohn, J. Krynen, S. Tedman-Jones, M.S. Dargusch, Promoting the columnar to equiaxed transition and grain refinement of titanium alloys during additive manufacturing, Acta Mater. 168 (2019) 261–274, https://doi.org/10.1016/j.actamat.2019.02.020.
34. Y.X. Jian, Z.F. Huang, Y. Wang, T.H. Wu, J.D. Xing, Solidification process, microstructure, and mechanical properties of Cr-doped high boron Fe-based alloy, Mater. Chem. Phys. 267 (2021) 124592, https://doi.org/10.1016/j.matchemphys.2021.124592.
35. E.B. Fonseca, A.G. Gabriel, L.C. Araújo, P.L.L. Santos, K.N. Campo, E.S.N. Lopes, Assessment of laser power and scan speed influence on microstructural features and consolidation of AISI H13 tool steel processed by additive manufacturing, Addit. Manuf. 34 (2020) 101250, https://doi.org/10.1016/j.addma.2020.101250.
36. Y.J Wen, B.C. Zhang, R.L. Narayan, P. Wang, X. Song, H. Zha, U. Ramamurty, X.H. Qu, Laser powder bed fusion of compositionally graded CoCrMo-Inconel 718, Addit. Manuf. 40 (2021) 101926, https://doi.org/10.1016/j.addma.2021.101926.
37. H.F Lu, Z. Wang, J. Cai, X. Xu, K.Y. Luo, L.J. Wu, J.Z. Lu, Effects of laser shock peening on the hot corrosion behaviour of the selective laser melted Ti6Al4V titanium alloy, Corro. Sci. 188 (2021) 109558, https://doi.org/10.1016/j.corsci.2021.109558.
38. Y.Y. Su, Z.F. Wang, H.F. Lu, K.Y. Luo, J.Z. Lu, Improved wear resistance of directed energy deposited Fe-Ni-Cr alloy via closed-loop controlling laser power, J. Manuf. Process. 75 (2022) 802–813, https://doi.org/10.1016/j.jmapro.2022.01.047.
39. Q.K. Shen, X.D. Kong, X.Z Chen, Signifcant transitions of microstructure and mechanical properties in additively manufactured Al–Co–Cr–Fe–Ni high-entropy alloy under heat treatment, Mater. Sci. Eng. A. 815 (2021) 141257, https://doi.org/10.1016/j.msea.2021.141257.
40. B. Zhang, S.Y. Liu, Y.C. Shin, In-Process monitoring of porosity during laser additive manufacturing process, Addit. Manuf. 28 (2019) 497–505, https://doi.org/10.1016/j.addma.2019.05.030.
41. F.L. Vecchiato, H.D. Winton, P.A. Hooper, M.R. Wenman, Melt pool microstructure and morphology from single exposures in laser powder bed fusion of 316L stainless steel, Addit. Manuf. 36 (2020) 101401, https://doi.org/10.1016/j.addma.2020.101401.

Chapter 4
Effect of the Melt Pool In-Situ Control on the Mechanical Properties of Laser Directed Deposition 316L Stainless Steel Layers

4.1 Introduction

Sliding wear and tensile property are key factors that limits the lifetime of metallic components for a wide range of functional and structural applications [1]. Therefore, the design of materials with perfect wear resistance and tensile property are in great demand to ensure the durability, reliability, and efficiency of metal contacts in harsh environments [2, 3]. 45 steel is commonly used in the manufacture of key components of agricultural machinery such as crankshafts and gears, which are prone to performance problems such as non-wear resistance and corrosion under typical working conditions. Fe-Ni-Cr alloy is often used for 45 steel coatings to improve mechanical properties, which often fabricated by directed energy deposition (DED) technology [4–7].

However, Fe-Ni-Cr alloy fabricated by traditional DED method is challenged by susceptibility to various problems induced by the high heat input during deposition. The wear resistance of the material is significantly affected by heat accumulation in the melt pool, which has hampered meeting the increasing demand of Fe-Ni-Cr alloy in the industries [8–10]. Luo et al. [11] developed a facile laser surface remelting-based surface treatment technique to fabricate heterogeneous gradient nanostructured metals, which significantly improved the microhardness value and wear resistance. Rahman et al. [12] used repetitive laser thermal cycling method to design and evaluate of thermally stable high temperature wear resistant alloys, which showed a better high temperature wear resistance. Zhu et al. [13] obtained the significant wear resistance by adjusting the interfacial structure.

The high-energy laser beam melts the substrate and powder to form the melt pool, and the heat accumulation problem can be solved by the effective monitoring and control of the melt pool state. Thermal image is usually neglected in the past sensing systems, which can be regarded as one of the most informative features of thermophysical dynamics [14–16]. He et al. [17] documented an investigation into

K. Luo et al., *Melt Pool In-Situ Controlling of Laser Additive Manufacturing Technology*, https://doi.org/10.1007/978-981-95-4597-1_4

the microstructures and melt pool geometry features of Ti6Al4V processed by selective laser melting (SLM). Lane et al. [18] described the use of reflectometer-based instrument and gave examples of the use of dynamic laser absorption measurements, in order to inform the interpretation of physical properties and process monitoring signals. Krauss et al. [19] used layered monitoring of the temperature distribution to gather information about the process stability and final part quality. Most of the researches on the thermal images concentrated on capturing the surface temperature of the deposition, but rarely studied to achieve the in-situ monitoring of melt pool width [20, 21].

Compared to the conventional fixed parameters, the closed-loop control system based on the proportion integration differentiation (PID) algorithm allows adaptive adjustment of process parameters to the part structure [22, 23]. Sain et al. [24] used one-dimensional input space to model each fuzzy component, which was then combined to provide a complete PID action. Zhong et al. [25] proposed a feedback model and the layerwise feedback control strategy and ensured the quality of large-scale parts. These studies have used PID control to achieve spatial modeling and quality control, but the effects of closed-loop control of MPS on the geometric profile of the melt pool, microstructure and wear resistance of the part have not been systematically investigated.

In this chapter, the wear and tensile properties were enhanced by closed-loop controlling of laser power and melt pool width. In the DED process for Fe-Ni-Cr alloys, the laser power was adjusted by PID algorithm to achieve real-time closed-loop control of the melt pool width. The effects of closed-loop control on the wear and tensile properties were investigated in both traditional mode (constant laser power) and control mode (closed-loop control of laser power). The results showed that superior wear resistance and better strength and toughness could be obtained in the control mode.

4.2 Experiment and Characterizations

4.2.1 LDED and Melt Pool In-Situ Control Process

The composition of the DED system was shown in Fig. 4.1. In this study, the DED unit of work contained a fiber-coupled diode laser system, gas-borne powder feeder, 3D printing head and a Vacuum chamber. Especially, the 3D printing head was equipped with a coaxial infrared camera for monitoring the real-time melt pool image. The additive manufacturing of the Fe-Ni-Cr alloy (chemical composition wt%: Cr 16–18, Ni 12–13, Mo 2–3, Si 0.8, Mn 0.2, C 0.03, P 0.03, S 0.03, O 0.03, Fe Bal.) was performed on 45 steel flat plate with dimensions of 150 mm × 150 mm × 10 mm (width ×length ×thickness).

In this study, 1000–2000 W were selected as the laser power reference groups. Through literature reading and orthogonal experiments, it was found that the

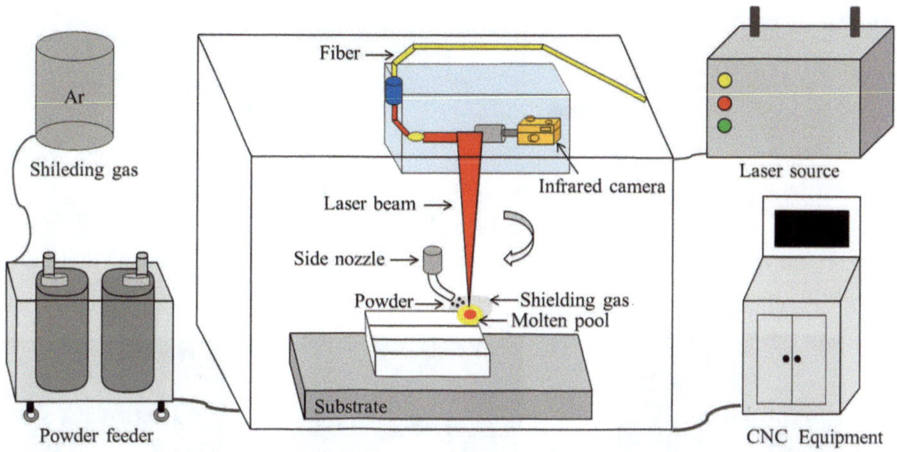

Fig. 4.1 The directed energy deposition system with in-situ closed-loop control of melt pool size

Table 4.1 The processing parameters in directed energy deposition

Parameters	Value	Units
Scanning speed	10	mm/s
Powder feed rate	18	g/min
Spot diameter	3	mm
Overlapping rate	50%	/

highest specimen binding, dilution rate and parameter optimization accuracy were achieved at 1600 W power, so 1600 W was preferably selected as the initial power for this study. Control groups between traditional mode and control mode were set to investigate the effect of closed-loop controlled MPS on microstructure and wear resistance. The deposition parameters used for two control groups were shown in Table 4.1.

Infrared image processing was the first important step in the closed-loop control process. The wavelength of the IR camera was 1064 nm. Images were captured at a resolution of 64 × 48 pixels, and recorded at a rate of 1000 fps for the entire build duration. The infrared imaging device was used for temperature calibration, and the data from the calibration was fit by the least square method. Figure 4.2 showed the process of extracting the MPS width from an infrared image. There were six main steps during image processing: (1) original melt pool image acquisition (Fig. 4.2a), (2) grey-scale processing (Fig. 4.2b), (3) denoising process used mean filtering method (Fig. 4.2c), (4) appropriate region of interest (ROI) selected by the threshold segmentation method (Fig. 4.2d), (5) smoothing by open operation (Fig. 4.2e), and (6) melt pool width acquisition (Fig. 4.2f) [26]. After optical focusing was completed, the spatial resolution of the original image was set to 0.15 mm/pixel. Combining the radiation temperature gradient of the melt pool and the software calculated pixel values, the MPS could be calculated [27, 28].

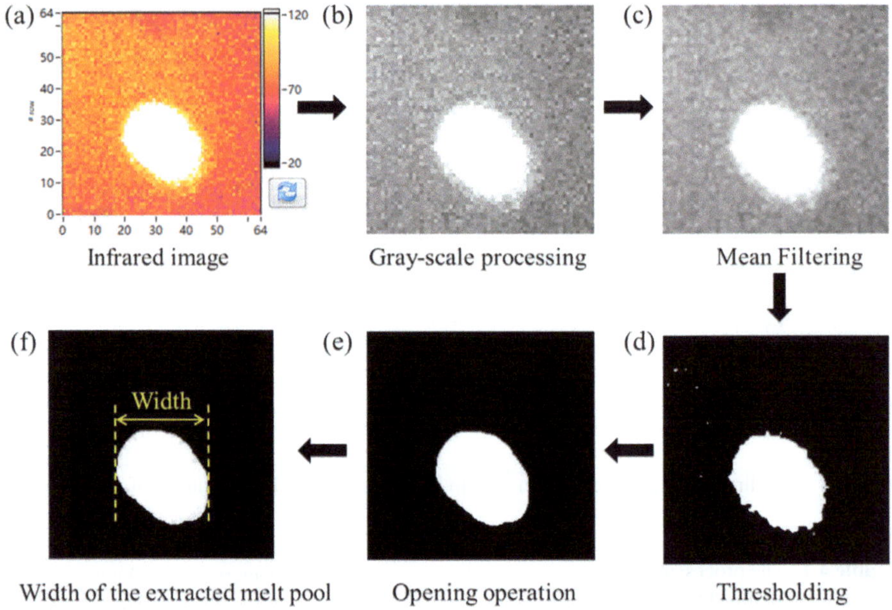

Fig. 4.2 The process of extracting the width of MPS from the infrared images (**a**) Infrared image, (**b**) gray-scale processing, (**c**) mean filtering, (**d**) thresholding, (**e**) opening operation, (**f**) image of melt pool width

When the laser and powder feeder were switched on, the high energy laser beam melted the substrate and powder, and then formed the first melt pool on the substrate. It was definite that the state of melt pool formation was highly transient, but through slowing down the rate of the melt pool image taken by the camera, it could be seen that the camera captured a gradually increasing circular pattern during the formation of the melt pool. When the circular pattern grew to the maximum, this transient picture could be selected, and then the first melt pool width after image processing was obtained. In order to evaluate the stability of the melt pool width, the first melt pool width obtained by camera during the deposition process was used as a reference value, the increase of each subsequent width value collected in real time was set as an input to the feedback control system, and the laser power was adjusted in real time by PID algorithm. The cycle time of the PID controller was 1 ms. It included the following steps. Firstly, the increase of each subsequent width value was defined as

$$e(k) = \frac{W_r(k) - W_1(k)}{W_1(k)} \tag{4.1}$$

where $W_1(k)$ was the first detected width during deposition, $W_r(k)$ was the subsequent MPS in real-time, and e(k) was the error between them. The positive error indicated that the real-time MPS was too small and the laser power needed to be increased, while the negative error indicated the opposite. Secondly, the change of self-correcting laser power was defined as

$$\Delta P = K_P \left\{ e(k) + \frac{1}{T_I} \sum_{m=1}^{k} e(m) + T_D \left[e(k) - e(k-1) \right] \right\} \tag{4.2}$$

where the K_p was the proportional gain, T_I was the integral time, T_D was the differential time. The proportional, integral and differential parts of the PID function coordinately controlled the laser power, which made the system have better dynamic performance [25]. In addition, to address the interference of signal noise on the derivative control, a low pass filter was used to filter out most of the noise. The signal strength was increased by adjusting the sensor gain value and the upper integration time limit was set to 800 μs. Through some exploration experiments, and the dynamic balance relationship between the laser power and the error of MPS, the PID parameters in this study were set as follows: K_P was 200, T_I was 0.4 *and* T_D was 0.5 [29, 30]. The adaptive adjustment of the laser power allowed the MPS width to be controlled within the optimum range, thus improving the quality of the part.

4.2.2 Tensile Test

The deposition samples and tensile test process pictures were shown in Fig. 4.3. In the experiments, single track coupon (Fig. 4.3a) and single layer coupon (Fig. 4.3b) specimens were deposited. Tensile specimens were taken from single-layer deposited samples at later stages along the building direction. The tensile gauge was parallel to the direction deposition, and the material in the tensile gauge length was 100% deposited material. In addition, each experiment was carried out by three parallel specimens. A double column electronic universal testing machine with the video extensometer was chose for tensile tests. The dimension of the tensile specimen was shown in Fig. 4.3e, where the shaded region indicated the specimen thickness of 2 mm.

4.2.3 Wear Test

Figure 4.4 showed the physical view of the tribological test set-up and configuration of ball-on-flat tribometer. HT-1000 universal friction and wear apparatus was composed of four parts: computer control system, high temperature furnace, mechanical friction system, and loading mechanism. The friction experiments were performed at room temperature. To eliminate any surface roughness effects, all specimens were mirror polished to an average surface roughness of 0.05 μm. A Si_3N_4 ball (~2200 HV) with a diameter of 6 mm was used as an oscillating counter-body, thus concentrating

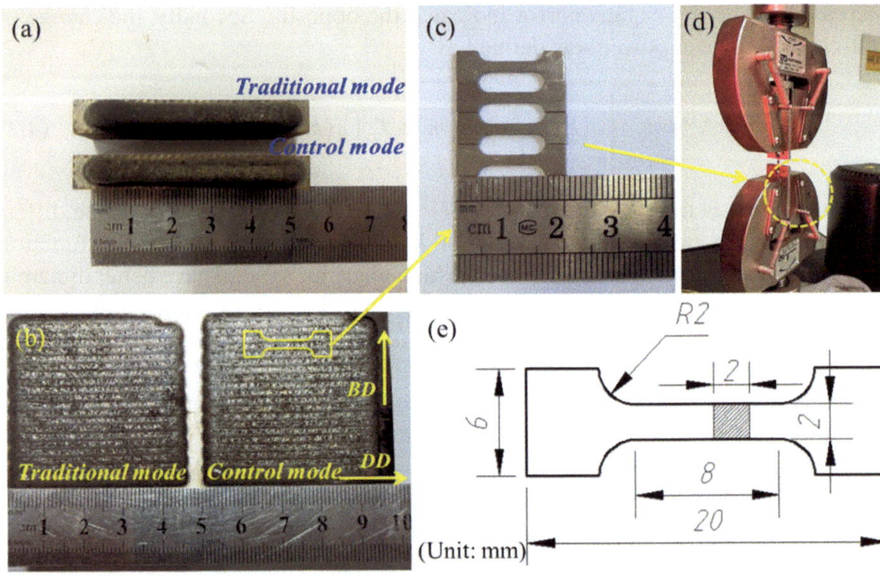

Fig. 4.3 The deposition samples and tensile test process pictures. (**a**) Single track coupon, (**b**) single layer coupon, (**c**) tensile specimens, (**d**) partial enlargement picture of the tensile process, and (**e**) dimensions of the tensile specimen

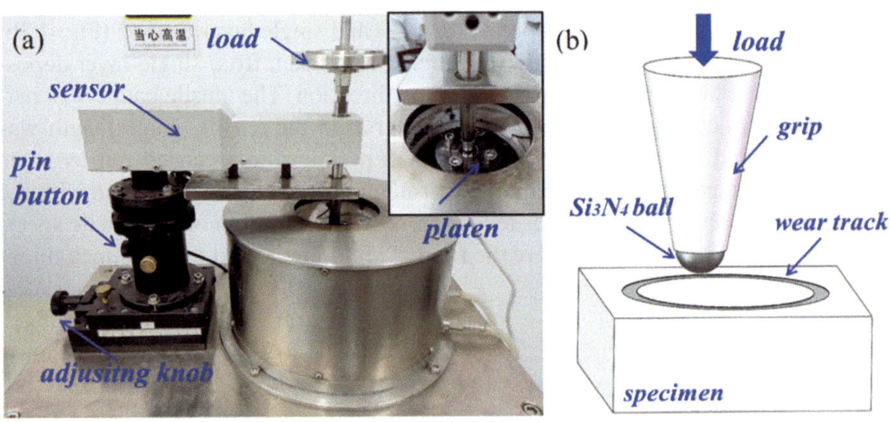

Fig. 4.4 Physical diagram of wear equipment and schematic diagram of reciprocating tribometer

all wear to the DEDed Fe-Ni-Cr specimen. The wear tests were performed at a relative humidity of about 50%, an average sliding speed of 0.12 m/s, a load of 1000 N and for a total sliding distance of 144 m.

At least three parallel specimens were used for each case, and the specimens were ultrasonically cleaned in ethanol and dried in air prior to the dry-sliding wear test. During the test, the coefficient of friction was recorded continuously and the average steady-state value was calculated. After the test, the wear scars and the wear volume were measured using a 3D optical measurement system Keyence VK-X250k. The wear rate k was calculated using Eq $k = \frac{V}{F_N \cdot s}\left[\mathrm{mm}^3/\mathrm{Nm}\right]$, taking into account the wear volume V, the load F_N and the sliding distance s [31, 32].

In addition, the microhardness distribution of the deposited layer along the deposition direction was measured using an HVS-1000 Vickers microhardness tester, with the load of 5 N and the dwell time of 10 s.

4.3 Tensile Property

4.3.1 Engineering Stress-Strain Curves

Figure 4.5 shows the tensile stress-strain curves of single group and the histogram with errors for all parallel specimens. As can be seen in Fig. 4.5a, the ultimate tensile strength of three samples under the traditional mode are 673.12 MPa, 1008.35 MPa and 908.12 MPa while the ultimate tensile strength under the control mode are 1070.01 MPa, 938.6 MPa, and 770.79 MPa, respectively. The elongation under the traditional mode are 2.3%, 1.6%, and 1.2% while the elongation under the control mode are 1.9%, 1.3%, and 0.8%, respectively. It can be found that the ultimate tensile strength (UTS) of the specimens increased by ~59% at 1600 W, while the UTS decreased by ~6.9% and ~15.2% at 1800 W and 2000 W, respectively. In addition, elongation is decreased at each power.

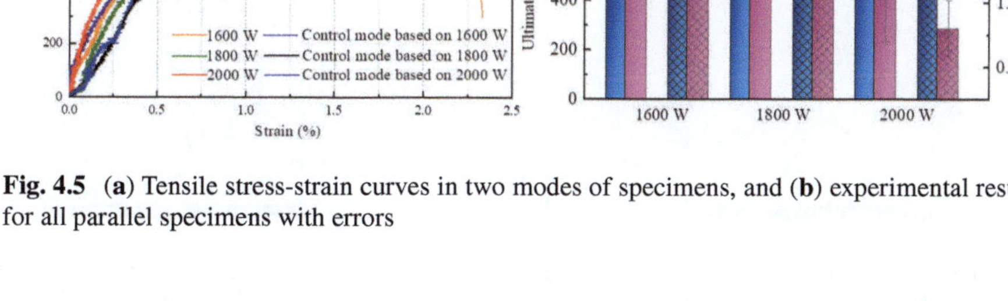

Fig. 4.5 (**a**) Tensile stress-strain curves in two modes of specimens, and (**b**) experimental results for all parallel specimens with errors

4.3.2 Fractural Morphology Analysis

The fracture morphologies of specimens under two modes are shown in Fig. 4.6. Cracks tend to develop where stresses are concentrated. When the laser power is1600 W, the crack source is found in the central region of the fracture and is accompanied by the river pattern, which is typical of cleavage fractures in the traditional mode (Fig. 4.6a). As shown in Fig. 4.6a′, densely distributed dimples with the size ranges from 3 μm to 14 μm appear on the cleavage plane. Prabhakaran and Kalainathan [31] believed that the dense distribution of dimples led to the higher elongation. The mechanism is a mixture of intergranular brittle and ductile fracture. In the control mode (Fig. 4.6b), the crack source appears at the convergence of the radial stripe, and the fracture path develops along the grain boundary which mainly presents brittle thin layer splitting shape. The sizes of the cleavage planes on one side of the radiating stripe is in the range of ~36 to ~71 μm (Fig. 4.6b′).

At 1800 W, the crack source which presents as radiating cleavage plane is also found in the middle region of the fracture in the traditional mode (Fig. 4.6c). A magnified observation of the central region reveals that the smallest cleavage plane is ~106.5 μm (Fig. 4.6c′). However, the crack source appeared on one side of the fracture in the control mode (Fig. 4.6d). The fracture morphology is

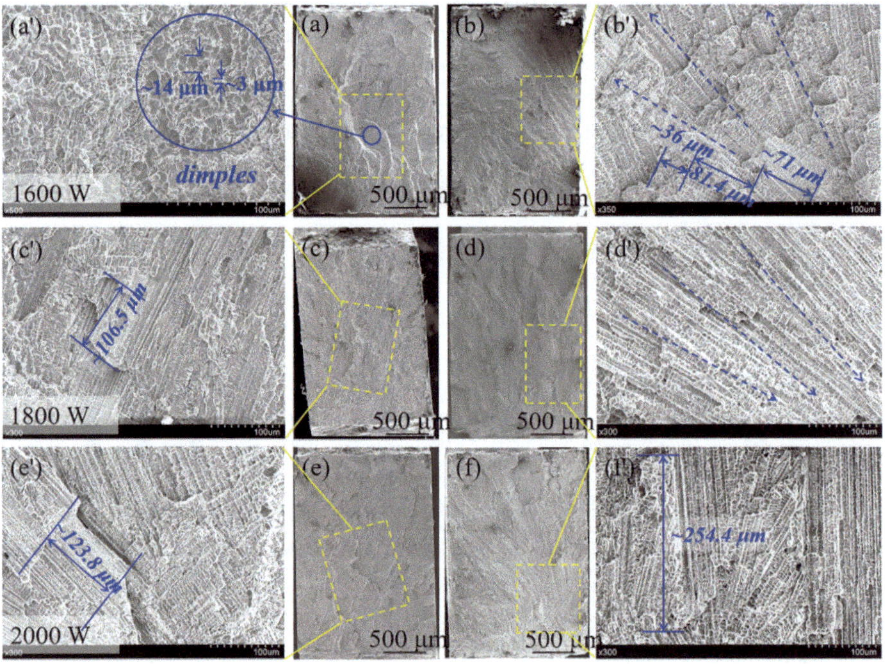

Fig. 4.6 SEM images of fracture morphology in the traditional mode (**a**, **c**, **e**) and the control mode (**b**, **d**, **f**). (**a**, **b**) and (**a′** and **b′**) 1600 W, (**c** and **d**) and (**c′** and **d′**) 1800 W, (**e** and **f**) and (**e′** and **f′**) 2000 W

dominated by radiating cleavage planes which are larger than 300 μm (Fig. 4.6d′). When the laser power is 2000 W, the crack sources appear on the fracture side in both modes (Fig. 4.6e, f), the main difference being that the size of the cleavage plane increases from ~123.8 μm in the traditional mode to ~254.4 μm in the control mode. The fracture morphology significantly affected the tensile property [32]. In the control mode, the UTS value is significantly increased due to the reduction in the size of cleavage planes based on 1600 W; conversely, due to the increase in the size of the cleavage planes based on 1800 W and 2000 W, the UTS value is reduced.

The EDS analysis of the fracture morphology at different power levels up to 2000 times magnification is shown in Fig. 4.7. At 1600 W, the EDS elemental examination of the dimples reveals that the elements are predominant as Fe and Cr (Fig. 4.7a). In the control mode (Fig. 4.7b), the dendritic second phase are observed at the fracture path along the grain boundary. Li et al. [33] believed that this was the main cause of the intergranular brittle fracture. From the EDS analysis, trace elements such as Si and Mo are detected at the fractures, which precipitate at the grain boundary and make the grain boundary brittle. Similarly, at laser powers of 1800 W and 2000 W, Fe and Cr are detected as the major elements in the traditional mode (Fig. 4.7c–e), while precipitation of trace elements (including Si and Mo) is detected in the control mode (Fig. 4.7d–f). In particular, the highest content of trace elements is precipitated at 1600 W in the control mode. Therefore, the number and sizes of the cleavage planes during the stretching process and the precipitation of trace elements at the fracture are the two main causes of the variation in elongation and tensile strength of the specimens.

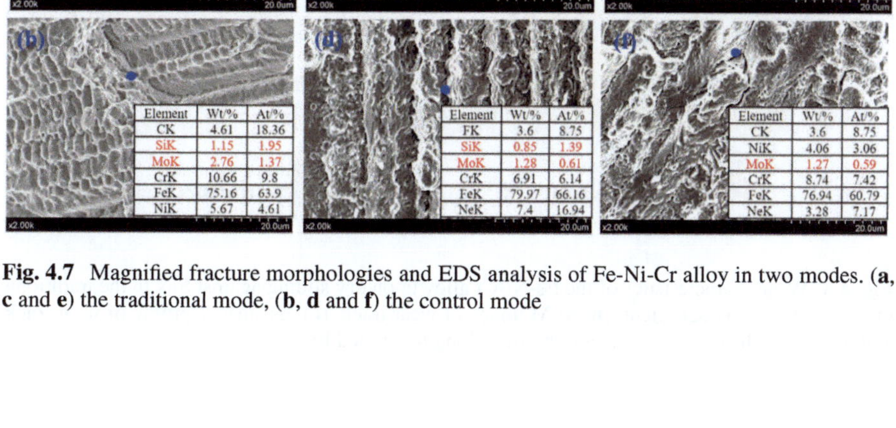

Fig. 4.7 Magnified fracture morphologies and EDS analysis of Fe-Ni-Cr alloy in two modes. (**a**, **c** and **e**) the traditional mode, (**b**, **d** and **f**) the control mode

4.4 Wear Property

4.4.1 Wear Resistance of the Traditional Mode

The worn surface of the specimen exhibits the phenomenon of flaking, mainly in the form of metal debris near the inner side of the wear ring through SEM observation (Fig. 4.8a). The largest of the debris reaches a width of ~174 μm. Upon wear exposure at room time, the coefficient of friction (CoF) of Fe-Ni-Cr alloy at steady state decreases to be 0.6 ± 0.02 in the traditional mode (Fig. 4.8b). By the three-dimensional morphology of the wear track (Fig. 4.8c), it can be observed that the pits are distributed in the wear ring and their depths range from −9 to 9 μm. From the analysis of the surface depth profile (at the blue dashed line), some significant bulges are found in the pit depth characterization, ranging from 1–6 μm in height (Fig. 4.8d). It is considered as the microcracks when the heights are less than 3 μm, and other larger ones are considered as large cracks. The maximum depth of the wear pits reaches −3.23 μm, which indicates severe damage of specimen in the traditional mode.

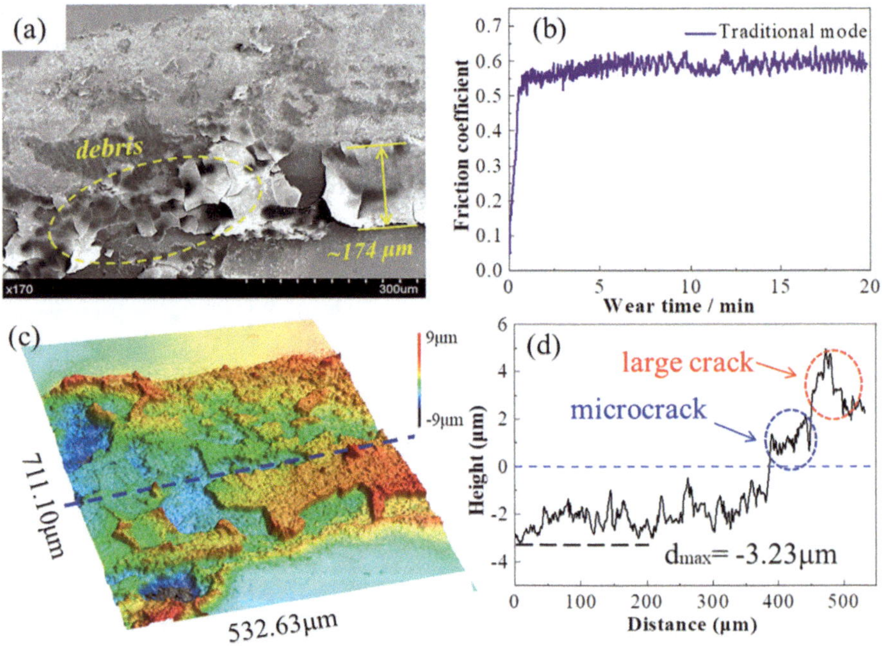

Fig. 4.8 Wear characteristics of the Fe-Ni-Cr alloy upon dry sliding against SiN in the traditional mode. (**a**) Friction coefficient, (**b**) SEM image of wear track, (**c**) 3D surface profile of wear track, and (**d**) pit depth curve of wear morphology along the dotted line

Figure 4.9 illustrates the wear morphology of specimens in the traditional mode. As can be seen in Fig. 4.9a, the friction ring shows two significantly different states as deposition proceeds. In the early stage of deposition (Fig. 4.9b), the specimen shows typical sliding wear. The SEM image at higher magnification shows the adhesion of small particles on the surface of the specimen after the wear test. In the later stage of the deposition (Fig. 4.9c), metal adhesion occurs when the Si_3N_4 balls are rubbed against the specimen surface. The broken adhesion during the rubbing process leads to the phenomenon of flaking and the abrasion of the specimen, which represents the typical adhesive wear mechanism [11, 34].

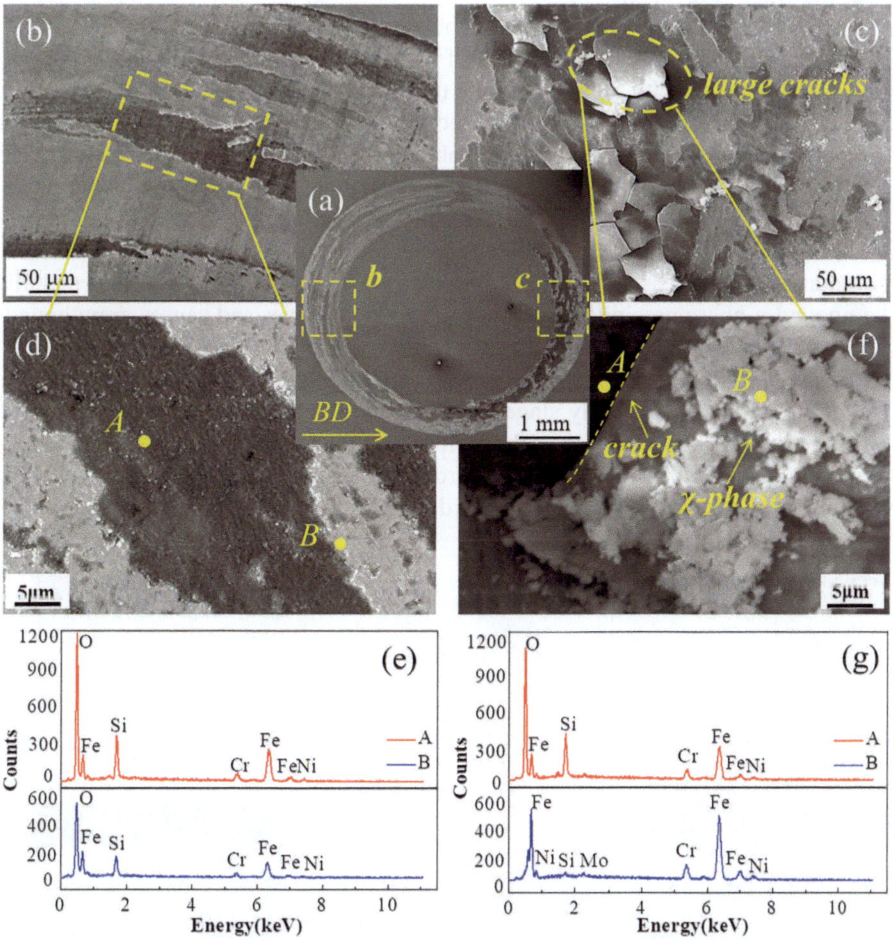

Fig. 4.9 SEM images of wear morphologies and point scanning results of Fe-Ni-Cr alloy in the traditional mode. (**a**) The complete wear pattern in the traditional mode, (**b**) SEM image of the wear pattern in the early stage of deposition, (**c**) SEM image of the wear pattern in the later stage of deposition, (**d**) magnified SEM image in the dashed line in (**b**), (**e**) EDS analysis for A and B in (**d**), (**f**) magnified SEM image in the dashed line in (**c**), and (**g**) EDS analysis for A and B in (**f**)

As shown in Fig. 4.9d–g, there are SEM images of wear morphology and EDS results focusing on the dominant wear mechanism in the traditional mode. Among them, Fig. 4.9d, f respectively correspond to the SEM magnified images in the early and late stages of the deposition process. From EDS analysis of the surface (point A) and the adhering material (point B) in Fig. 4.9d, it can be found that more oxygen and Si elements are detected in the area where spalling of the specimen occurs. It indicates that a certain degree of damage to the Si_3N_4 occurs during the sliding friction process. In a later stage of deposition, EDS analysis of the wear surface at the cracked metal fragment (point A) and the adhering particles (point B) reveals that high levels of oxygen and silicon elements are also detected on the wear surface near the crack (Fig. 4.9f).

In addition to the specimen and the abrasive material, due to the precipitation of χ-phase at the crack, a certain amount of Mo element is found at the adherent particles. The χ-phase with 58 atoms in the unit cell belongs to the TCP phase family. It exhibits relatively high hardness at room temperature and high thermodynamic stability. TCP phases are mostly undesirable in metallic materials, such as Fe-Ni-Cr alloy, as they deteriorate the mechanical (e.g. toughness) and physicochemical (e.g. corrosion resistance) properties [35, 36]. In the later stages of the deposition process, the thermal accumulation effect causes the increasing melt pool width, and simultaneously increases the tendency for χ-phase precipitation. The intermetallic compound χ-phase precipitates along the grain boundaries and causes grain boundary embrittlement, resulting in cracks at last.

Figure 4.10 illustrates the cross-sectional microstructure of the wear tracks in the traditional mode. In the early stage of deposition, the wear tracks are observed to appear vertically above the lap of the melt pool (Fig. 4.10a). Microscopic characterization of the A_1 region reveals that the growth of planar and dendritic crystals at the bottom of the melt pool is oriented toward the center of the respective melt pool due

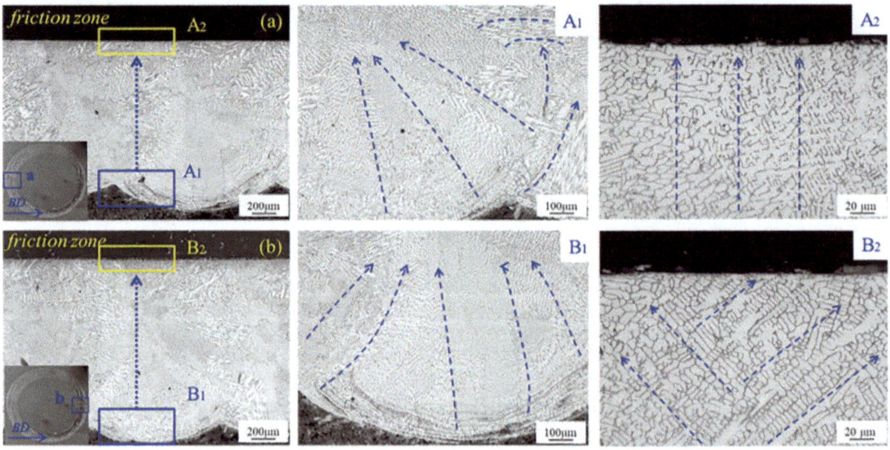

Fig. 4.10 Microstructure of the cross section under the wear tracks of the specimen in the traditional mode. (**a**) The early stage of deposition process, and (**b**) the late stage of deposition process

to the overlap of the two pools. The microstructure in the middle of the melt lap is oriented vertically towards the top of the melt pool and is dominated by small-sized cell crystals (Fig. 4.10B_2). In the later stages of deposition, wear tracks are observed to appear directly above the bottom center of the melt pool, which is different from the early stages of deposition. Characterization of the B_1 region reveals that the microstructure grows toward the center of the melt pool and cellular structure with different growth directions appears at the top of the melt pool. As shown in the B_2 region, the two growth directions of the microstructure show an angle of 90°.

In the traditional mode, the wear resistances of wear region in two stages are closely related to the microstructure and grain orientation of the cross-section. The relative positions of the melt pool and the wear scar as well as the dynamic thermal effects determine the microstructure. The microstructure at the melt pool lap has a unique crystallographic texture without any preferred orientation. The wear rate (i.e., material loss) in the early stages of deposition is ~2.0×10^{-5} mm^3/Nm. While the wear rate increases significantly in the later stage of deposition, reaching ~3.5×10^{-5} mm^3/Nm. Thermal accumulation effects and the increase in melt pool size lead to the increase of austenite content and a typical cellular substructure at the microscopic level, resulting in strong properties anisotropy. During the wear process, the cellular structure boundary is broken into particles, forming fine precipitates as well as large amounts of metal debris on the specimen surface. In this case, defects and inclusions representing the cutting edges promote counter-body wear and material transfer. The increase in the density of cracked defects leads to the decrease in wear resistance [37].

4.4.2 Wear Resistance of the Control Mode

Figure 4.11 shows the wear characteristics of the Fe-Ni-Cr alloy upon dry sliding against Si_3N_4 in the control mode. SEM observation shows uniform wear tracks on the specimen surface along the sliding direction in Fig. 4.11a. The steady-state COF is 0.48 ± 0.02 when the wear is performed at room temperature (Fig. 4.11b), which is an improvement in wear resistance compared to that in the traditional mode. By the three-dimensional morphology (Fig. 4.11c), it can be seen that the wear tracks are much even after control. Most of the pits are also observed in the middle of the wear ring, and their depths range from −3 to 2 μm. From the analysis of the surface depth profile near the inside of the wear ring (at the blue dashed line), it is clear that the uniform depth of the wear pits is −0.13 μm (Fig. 4.11d).

Figure 4.12 illustrates the wear morphology of specimens in the control mode. Same as Fig. 4.9a, the friction ring shows two significantly different states as deposition proceeds in Fig. 4.12a. Some microcracks and debris are still observed on the surface of friction ring (Fig. 4.12b). The adhesion of the metal indicates the adhesive wear in the early stage of deposition. In the later stage of the deposition, some long and shallow grooves running along the sliding direction with high density are found in Fig. 4.12c, confirming that the dominant wear mechanism of the Si_3N_4 ball

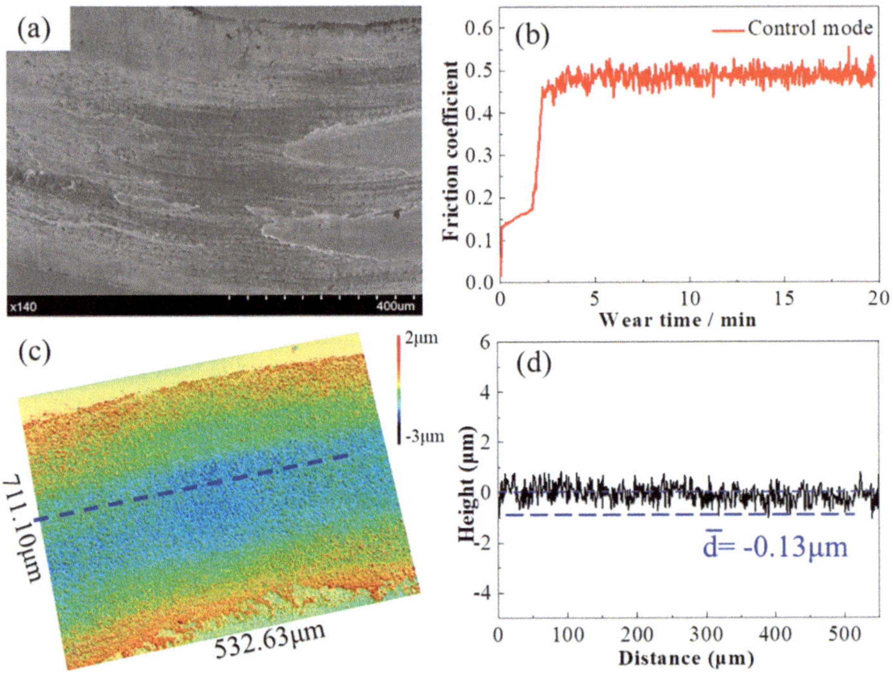

Fig. 4.11 Wear characteristics of the Fe-Ni-Cr alloy upon dry sliding against Si3N4 in the control mode. (**a**) Friction coefficient, (**b**) SEM image of wear track, (**c**) 3D surface profile of wear track, and (**d**) pit depth curve of wear morphology along the dotted line

is abrasive wear [38, 39]. In general, the wear mechanism is a typical mechanical mixing wear in the control mode.

As shown in Fig. 4.12e, the EDS analysis of the wear tracks reveals that more Fe and Mo are detected on the surface where there are many aggregations of adhered particles and crack defects (point A). This discovery is the same as that at the early stage in the traditional mode. In addition, more oxygen content is detected at the wear tracks (point B), indicating the presence of oxidation wear. In the late stage of deposition, the SEM images show wear tracks dominated by uniform grooves consistent along the sliding direction (Fig. 4.12f). The composition detected at the grooves is dominated by the elemental composition of the grinding material and the specimen, with the highest content of Fe-based elements (Fig. 4.12g).

Figure 4.13 illustrates the microstructure of the wear track cross-section in the control mode. In the early stage of deposition process (Fig. 4.13a), the wear tracks are found to be located directly above the center of the melt pool and the depth of the melt pool is significantly reduced compared to that in the traditional mode. Characterization of the bottom region of the melt pool (A_1) reveals that the organization mainly goes through three stages of planar crystal, dendritic crystal, and equiaxed crystal, getting equiaxed grain crystallization at 300 μm from the bottom. Further observation of the microstructure in the top region of the melt pool (A_2)

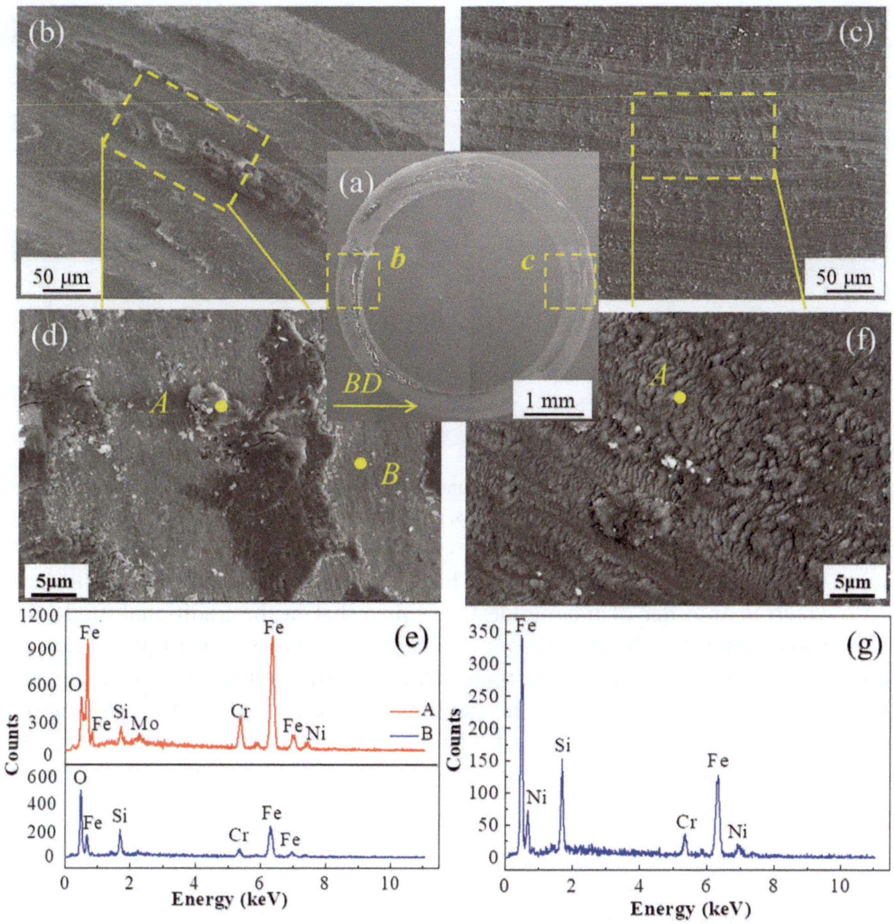

Fig. 4.12 SEM images of wear morphologies and point scanning results of Fe-Ni-Cr alloy in the control mode. (**a**) The complete wear pattern in the control mode, (**b**) SEM image of the wear pattern in the early stage of deposition, (**c**) SEM image of the wear pattern in the later stage of deposition, (**d**) magnified SEM image in the dashed line in (**b**), (**e**) EDS analysis for A and B in (**d**), (**f**) magnified SEM image in the dashed line in (**c**), (**g**) EDS analysis for A in (**f**)

reveals that the grain size of equiaxed crystal reaches ~30 μm and the wear rate is 1.8×10^{-5} mm^3/Nm.

In the later stage of deposition (Fig. 4.13b), the relative position of the wear tracks to the bottom of the melt pool is the same as that in the early stage. The difference is that the equiaxed grain crystallization is achieved at a distance of ~219 μm from the bottom, and the grain size of the equiaxial crystals on the near surface of the wear track is only ~15 μm. The closed-loop control of the laser power influences the cooling rate of the melt pool, which accelerates the process of grain refinement. The homogenization of the microstructure and minimization of the interfacial defects also eliminate the effect of thermal effects accumulation on the adhesive

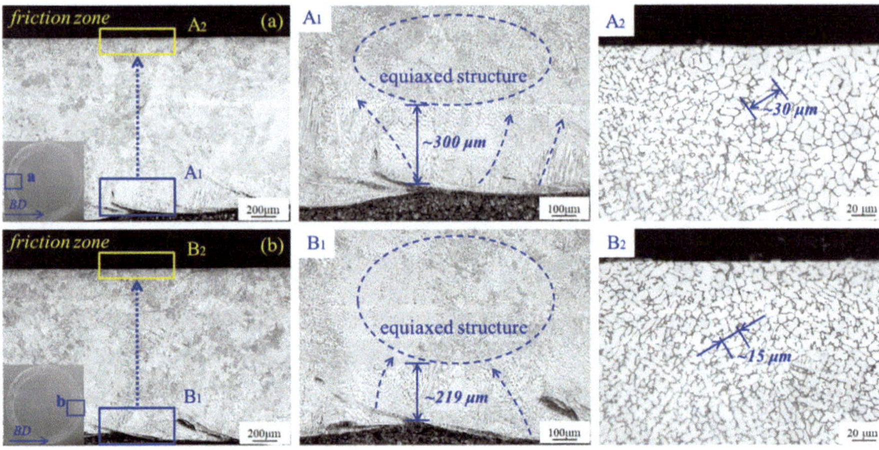

Fig. 4.13 Microstructure of the cross section under the wear tracks of the specimen in the control mode. (**a**) The early stage of deposition process, and (**b**) the late stage of deposition process

wear, forming dense equiaxed crystals. This results in a wear rate of ~3.0×10^{-6} mm^3/Nm, which is an-order-of magnitude lower than that in the traditional mode. The closed-loop control of the laser power effectively eliminates anisotropy, resulting in complete homogenization of the microstructure. The cellular structure and thin striped structures that appear in the traditional mode are replaced by an equiaxed crystal structure without any preferred orientation. The grain refinement further increases the microhardness of the material, which is another reason for the improved wear resistance in the control mode.

4.5 Influence Mechanisms of Melt Pool Control on the Mechanical Properties

4.5.1 Tensile Property

At 1600 W, schematic fracture mechanism of Ni-Cr alloy tensile specimens is shown in Fig. 4.14. There is a significant correspondence between grain size and fracture morphology in both modes. In the traditional mode, there are some distinct brittle fracture features such as cleavage planes and tear ridges are observed at the fracture. The grains at the fracture grow in a radial direction around the stress concentration point. The fracture path of the sample develops along the grain boundary, and the grains are mainly large-sized dendrites. The cleavage planes are characterized by big size and low density distribution. In addition, the dimple morphology is also observed on the cleavage plane. Saboori et al. [40] thought it was a mixture of intergranular brittle and ductile fracture.

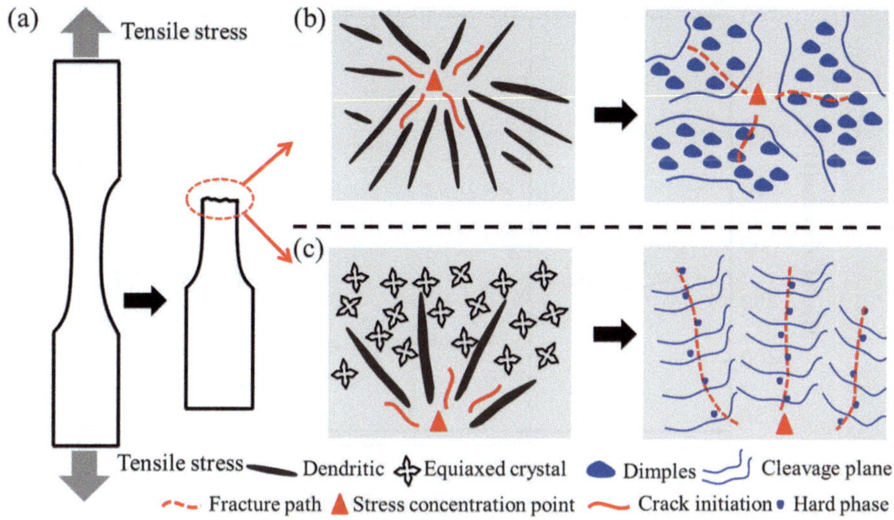

Fig. 4.14 Schematic fracture mechanism of tensile specimen fabricated by DED. (**a**) Fracture process of tensile specimens, (**b** and **c**) schematic fracture mechanism of tensile specimens in two modes

In the control mode, the melt width is effectively suppressed, leading to grain refinement and the formation of more grain boundaries. The grains grow to the same side along the stress concentration point, and the fracture path of the sample develops along the grain boundary. The accumulation of dislocations by the increased number of grain boundaries causes stress concentration and cracking, resulting in the formation of small-sized cleavage planes with a radially laminated distribution at the fracture. Sun et al. [41] claimed that this further enhances the brittleness of the specimen and is typical of intergranular brittle fracture. In addition, the high content of trace element detected in the fracture cracks is another reason for the increased tensile strength of the specimens, with the precipitation of trace elements at grain boundaries promoting the growth of hard phases. According to the Hall-Petch formula [42], the grain strength along the grain brittleness is higher, which makes the ultimate tensile strength of the material better.

4.5.2 Wear Property

The hardness values distributed over different areas of specimens are shown in Fig. 4.15. It is noteworthy that the hardness shows different results in two deposition modes. During the melted manufacturing process with constant laser power, the hardness becomes smaller with increasing scanning length. On the contrary, the hardness values gradually increase with the deposition time under the closed-loop control of laser power. Interestingly, in the A region of the cladding layer, i.e. at the

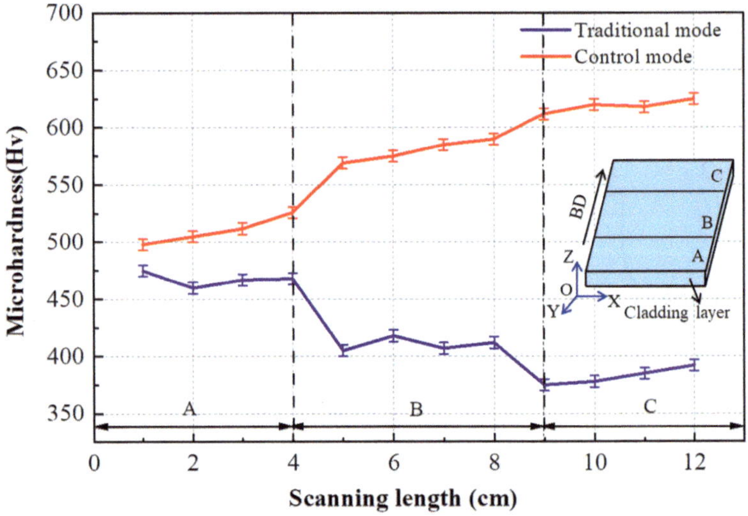

Fig. 4.15 Hardness values for three areas of the Fe-Ni-Cr alloy along the deposition direction in two modes

beginning of the deposition process, hardness values in both two modes range from 460–538 HV, due to the limited effect of the closed-loop control in the initial stage. Halfway through the deposition process, the difference in hardness values becomes more pronounced for the significant effect after control. The difference in the distribution of hardness values gets the most obvious effect in the later stages. In the control mode, the hardness reaches the maximum of 620 HV, while the lowest value in the traditional mode is only 375 HV [43].

Figure 4.16 illustrates the wear property and microstructure influence schematic of DEDed Fe-Ni-Cr alloy at closed-loop controlled laser power. The continuous thermal accumulation and slow cooling rate at constant laser power limit the transformation of ferrite to austenite, finally leading to more dendrite structures generation (Fig. 4.16a). However, due to the reduced thermal accumulation and accelerated cooling rate, the austenite further increases in the control mode (Fig. 4.16b). It limits the growth of large slate-like grains during super cooling solidification and finally leads to the formation of uniformly sized equiaxed crystals. Grain refinement strengthening significantly increases the microhardness of the material, resulting in a gradual increase in hardness values with the direction of deposition.

In the later stages of deposition, the thermal accumulation effect is mainly manifested in the traditional mode, while the thermal suppression effect is mainly manifested in the control mode. Therefore, the main difference of the wear performance in two modes is concentrated in the later stage of deposition. The microstructure differences in the cross-section of the wear tracks significantly affect the wear resistance of the material. Cellular structures with cross-growth in a 90° direction are observed in the traditional mode (Fig. 4.16c), while dense equiaxed crystals with no preferential directional growth are observed in the control mode (Fig. 4.16d). This

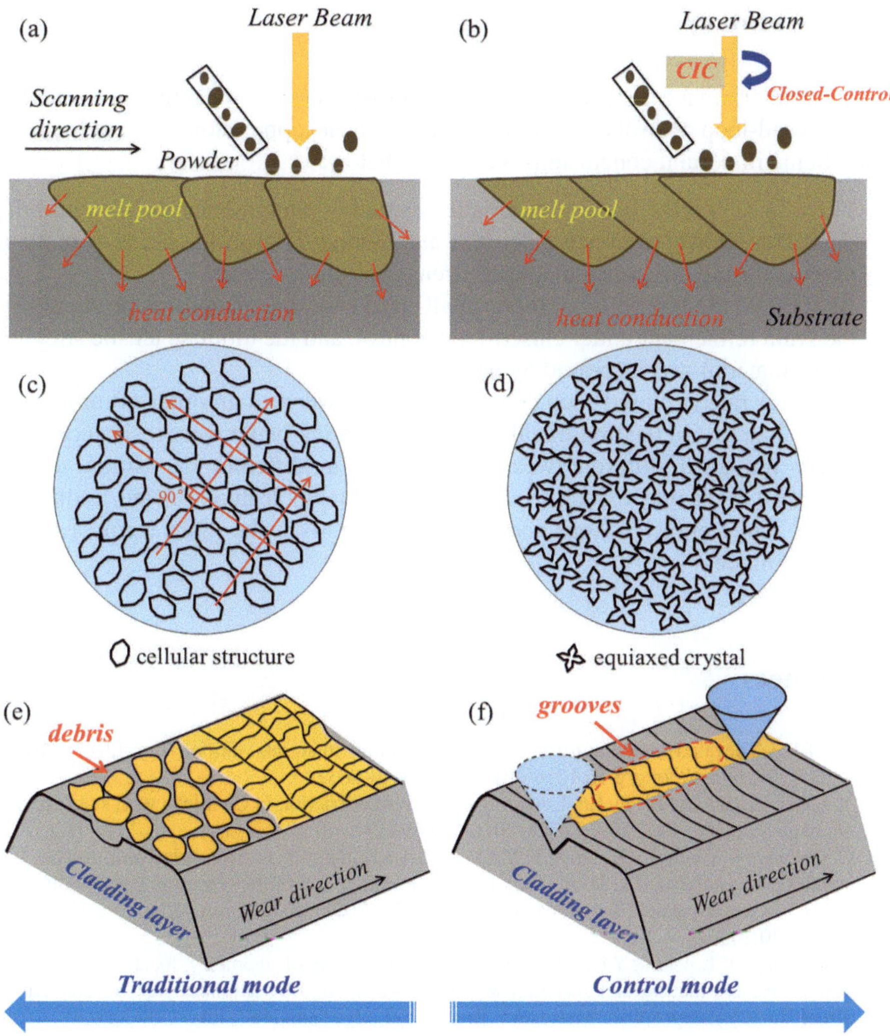

Fig. 4.16 Schematic diagram of the friction process and wear morphology in both modes (**a**) traditional LDED process, (**b**) closed-control process with a CIC system, (**c**) cellular structure, (**d**) equiaxed structure, (**e**) wear mechanism in the traditional mode, (**f**) wear mechanism in the control mode

also leads to the significant difference in the wear mechanism. In the traditional mode, due to the reduction of hardness, the oxide layer is severely damaged. The cellular structure boundary is broken into particles, forming fine precipitates as well as a large amount of metal debris on the specimen surface, which in this case is mainly dominated by adhesive and oxidative wear (Fig. 4.16e). In the control mode, uniform grooves which parallel to the sliding direction are observed on wear tracks, indicating the wear mechanism of abrasive wear (Fig. 4.16f).

4.6 Summary

In this chapter, wear resistance and tensile property were significantly improved due to the closed-loop control of laser power and the heat suppression effect of the melt pool. Some important conclusions were listed below:

1. There were two fracture characteristics including intergranular ductile and brittle fracture in the specimens. The size and number of cleavage plane were main reasons for the difference in tensile strength of the specimens.
2. At 1600 W, the formation of more small sized cleavage planes was promoted by the grain refinement under closed-loop control, and the ultimate tensile strength of the material was increased by up to ~59%.
3. The wear rate at closed-loop controlled laser power was an order of magnitude lower than that in the traditional mode, representing better wear resistance.
4. There remained different wear morphology in different deposition stages. Adhesive wear and oxidation wear were the main features in the traditional mode, while abrasive wear and adhesion wear were the main features in the control mode.

References

1. P.J. Blau, Elevated-temperature tribology of metallic materials, Tribol. Int. 43 (7) (2010) 1203-1208, https://doi.org/10.1016/j.triboint.2010.01.003.
2. X. Chen, Z. Han, X. Li, K. Lu, Lowering coefficient of friction in Cu alloys with stable gradient nanostructures, Sci. Adv. 2 (12) (2016) 1601942.
3. C.Y. Wang, W. Cheng, Y.K. Shao, K.Y. Luo, J.Z. Lu, Cavitation erosion behaviour of AISI 420 stainless steel subjected to laser shock peening as a function of the coverage layer in distilled water and water-particle solutions, Wear. 470–471 (2021) 203611, https://doi.org/10.1016/j.wear.2020.203611.
4. X.D. Nong, X.L. Zhou, Y.D. Wang, L.Yu, J.H. Li, Effects of geometry, location, and direction on microstructure and mechanical properties of 15–5PH stainless steel fabricated by directed energy deposition, Mater. Sci. Eng. A 821 (2021) 141587, https://doi.org/10.1016/j.msea.2021.141587.
5. J.Z. Lu, H.F. Lu, X. Xu, J.H. Yao, J. Cai, K.Y. Luo, High-performance integrated additive manufacturing with laser shock peening –induced microstructural evolution and improvement in mechanical properties of Ti6Al4V alloy components, Int. J. Mach. Tools Manuf. 148 (2020) 103475. https://doi.org/10.1016/j.ijmachtools.2019.103475.
6. T. Wang, Y.Y. Zhu, S.Q. Zhang, H.B. Tang, H.M. Wang, Grain morphology evolution behavior of titanium alloy components during directed energy deposition additive manufacturing, J. Alloys Compd. 632 (2015) 505–513, https://doi.org/10.1016/j.jallcom.2015.01.256.
7. A. Emelogu, M. Marufuzzaman, S.M. Thompson, N. Shamsaei, L. Bian, Additive manufacturing of biomedical implants: a feasibility assessment via supply-chain cost analysis, Addit. Manuf. 11 (2016) 97–113, https://doi.org/10.1016/j.addma.2016.04.006.
8. X. Xu, H.F. Lu, Y.Y. Su, M.X. Peng, F. Xing, K.Y. Luo, J.Z. Lu, Comparing corrosion behavior of additively manufactured Cr-rich stainless steel coating between conventional and extreme high-speed laser metal deposition, Corro. Sci. 195 (2022) 109976. https://doi.org/10.1016/j.corsci.2021.109976.

9. A. Vasinonta, J. Beuth, M. Griffith, Process maps for controlling residual stress and melt pool size in laser-based SFF processes, Proceedings of the Solid Freeform Fabrication Symposium (2000) 200–208.
10. L. Scime, J. Beuth, Melt pool geometry and morphology variability for the Inconel 718 alloy in a laser powder bed fusion additive manufacturing process, Addit. Manuf. 29 (2019) 100830, https://doi.org/10.1016/j.addma.2019.100830.
11. J.S. Luo, W.T. Sun, R.X. Duan, W.Q. Yang, K.C. Chan, F.Z. Ren, X.S. Yang, Laser surface treatment-introduced gradient nanostructured TiZrHfTaNb refractory high-entropy alloy with significantly enhanced wear resistance, J. Mater. Sci. Technol. 110 (2022) 43-56, https://doi.org/10.1016/j.jmst.2021.09.029.
12. N. U. Rahman, L. Capuano, S. Cabeza, M. Feinaeugle, A. Garcia-Junceda, M.B. de Rooij, D.T.A. Matthews, G. Walmag, I. Gibson, G.R.B.E. Römer, Directed energy deposition and characterization of high-carbon high speed steels, Addit. Manuf. 30 (2019) 100838, https://doi.org/10.1016/j.addma.2019.100838.
13. W.W. Zhu, C.C. Zhao, Y.W. Zhang, C.T. Kwok, J.H. Luan, Z.B. Jiao, F.Z. Ren, Achieving exceptional wear resistance in a compositionally complex alloy via tuning the interfacial structure and chemistry, Acta Mater. 188 (2020) 697-710, https://doi.org/10.1016/j.actamat.2020.02.039.
14. B.T. Gibson, Y.K. Bandari, B.S. Richardson, W.C. Henry, E.J. Vetland, T.W. Sundermann, L.J. Love, Melt pool size control through multiple closed-loop modalities in laser-wire directed energy deposition of Ti-6Al-4V, Addit. Manuf. 32 (2020) 100993, https://doi.org/10.1016/j.addma.2019.100993.
15. J. Haley, C. Leach, B. Jordan, R. Dehoff, V. Paquit, In-situ digital image correlation and thermal monitoring in directed energy deposition additive manufacturing, Optics. Express. 29 (2021) 9927-9941, https://doi.org/10.1364/OE.416659.
16. R.J. Williamsa, A. Piglioneb, T. Rønneberga, C. Jonesc, M. Phamb, C.M. Daviesa, P.A. Hooper, In situ thermography for laser powder bed fusion: Effects of layer temperature on porosity, microstructure and mechanical properties, Addit. Manuf. 30 (2019) 100880, https://doi.org/10.1016/j.addma.2019.100880.
17. Y.N. He, C. Montgomery, J. Beuth, B. Webler, Melt pool geometry and microstructure of Ti6Al4V with B additions processed by selective laser melting additive manufacturing, Mat. Des. 183 (2019) 108126, https://doi.org/10.1016/j.matdes.2019.108126.
18. B. Lane, I. Zhirnov, S. Mekhontsev, S. Grantham, R. Ricker, S. Rauniyar, K. Chou, Transient laser energy absorption, co-axial melt pool monitoring, and relationship to melt pool morphology, Addit. Manuf. 36 (2020) 101504, https://doi.org/10.1016/j.addma.2020.101504
19. H. Krauss, T. Zeugner, M.F. Zaeh, Layerwise monitoring of the selective laser melting process by thermography, Phys. Procedia. 56 (2014) 64–71, https://doi.org/10.1016/j.phpro.2014.08.097.
20. N. Boone, C. Zhu, C. Smith, I. Todd, J. Willmott, Thermal near infrared monitoring system for electron beam melting with emissivity tracking, Addit. Manuf. 22 (2018) 601–605, https://doi.org/10.1016/J.ADDMA.2018.06.004.
21. Z.Y. Zhang, Z.C. Liu, D.Z Wu, Prediction of melt pool temperature in directed energy deposition using machine learning, Addit. Manuf. 37 (2021) 101692, https://doi.org/10.1016/j.addma.2020.101692.
22. Z.R. Yan, W.W. Liu, Z.J. Tang, X.Y. Liu, N. Zhang, M.Z. Li, H.C. Zhang, Review on thermal analysis in laser-based additive manufacturing, Opt. Laser Technol. 106 (2018) 427–441, https://doi.org/10.1016/j.optlastec.2018.04.034.
23. L.P. Zheng, Q. Zhang, H.Z. Cao, W.J. Wu, H.L. Ma, X.P. Ding, J. Yang, X.M. Duan, S.Q. Fan, Melt pool boundary extraction and its width prediction from infrared images in selective laser melting, Mater. Des. 183 (2019) 108110, https://doi.org/10.1016/j.matdes.2019.108110.
24. D. Sain, B.M. Mohan, A simple approach to mathematical modelling of integer order and fractional order fuzzy PID controllers using one-dimensional input space and their experimental

realization, Journal of the Franklin Institute. 358 (2021) 3726-3756, https://doi.org/10.1016/j.jfranklin.2021.03.010.

25. Q. Zhong, X.Y. Tian, X.K. Huang, C.B. Huo, D.C. Li, Using feedback control of thermal history to improve quality consistency of parts fabricated via large-scale powder bed fusion, Addit. Manuf. 42 (2021) 101986, https://doi.org/10.1016/j.addma.2021.101986.
26. Q.Y. Lu, N.V. Nguyen, A.J.W. Hum, T. Tran, C.H. Wong, Optical in-situ monitoring and correlation of density and mechanical properties of stainless steel parts produced by selective laser melting process based on varied energy density, J. Mater. Process. Technol. 271 (2019) 520–531, https://doi.org/10.1016/j.jmatprotec.2019.04.026.
27. C. Luo, et al., Finite element analysis of temperature and stress fields during the selective laser melting process of thermoelectric SnTe, J. Mater. Process. Technol. 261 (2018) 74e85, https://doi.org/10.1016/j.jmatprotec.2018.06.001.
28. P.A. Hooper, Melt pool temperature and cooling rates in laser powder bed fusion, Addit. Manuf. 22 (2018) 548-559, https://doi.org/10.1016/j.addma.2018.05.032.
29. X.H. Zhou, D.L. Li, L. Zhang, Q.L. Duan, Application of an adaptive PID controller enhanced by a differential evolution algorithm for precise control of dissolved oxygen in recirculating aquaculture systems, Biosyst. Eng. 208 (2021) 186-198, https://doi.org/10.1016/j.biosystemseng.2021.05.019.
30. G.A.O. Adam, D. Zimmer, Design for additive manufacturing-element transitions and aggregated structures, CIRP J. Manuf. Sci. Technol. 7 (1) (2014) 20–28.
31. S. Prabhakaran, S. Kalainathan, Compound technology of manufacturing and multiple laser peening on microstructure and fatigue life of dual-phase spring steel, Mater. Sci. Eng. A. 674 (2016) 634-645, https://doi.org/10.1016/j.msea.2016.08.031.
32. Y.Y. Su, Z.F. Wang, J.C. Xie, G. Xu, F. Xing, K.Y. Luo, J.Z. Lu, Microstructures and mechanical properties of laser melting deposited Ti6Al4V/316L functional gradient materials, Mater. Sci. Eng. A. 817 (2021) 141355, https://doi.org/10.1016/j.msea.2021.141355.
33. Z. Li, J. Chen, S. Sui, C.L. Zhong, X.F. Lu, X. Lin, The microstructure evolution and tensile properties of Inconel 718 fabricated by high-deposition-rate laser directed energy deposition, Addit. Manuf. 31 (2020) 100941, https://doi.org/10.1016/j.addma.2019.100941.
34. J.K. Xiao, H. Tan, J. Chen, A. Martini, C. Zhang, Effect of carbon content on microstructure, hardness and wear resistance of $CoCrFeMnNiC_x$ high-entropy alloys, J. Alloy. Compd. 847 (2020) 156533, https://doi.org/10.1016/j.jallcom.2020.156533.
35. K. Wieczerzak, J. Michler, J.M. Wheeler, S. Lech, R. Chulist, Ł. Gondek, J. Czub, A. Hoser, N. Schell, P. Bała, An in situ and ex situ study of χ phase formation in a hypoeutectic Fe-based hardfacing alloy, Mater. Des. 188 (2020) 108438, https://doi.org/10.1016/j.matdes.2019.108438.
36. M. Pohl, O. Storz, T. Glogowski, Effect of intermetallic precipitations on the properties of duplex stainless steel, Mater. Charact. 58 (2007) 65–71, https://doi.org/10.1016/j.matchar.2006.03.015.
37. D. Kim, T. Kim, K. Ha, J.J. Oak, J.B. Jeon, Y. Park, W. Lee, Effect of heat treatment condition on microstructural and mechanical anisotropies of selective laser melted maraging 18Ni-300 steel, Metals 10 (2020) 410, https://doi.org/10.3390/met10030410.
38. P. Zhao, J. Li, Y. Zhang, X. Li, M.M. Xia, B.G. Yuan, Wear and high-temperature oxidation resistances of $AlNbTaZr_x$ high-entropy alloys coatings fabricated on Ti6Al4V by laser cladding, J. Alloy. Compd. 862 (2021) 158405, https://doi.org/10.1016/j.jallcom.2020.158405.
39. C.C. Zhao, Y.F. Zhou, X.L. Xing, S. Liu, X.J. Ren, Q.X Yang, Investigation on the relationship between NbC and wear-resistance of Fe matrix composite coatings with different C contents, Appl. Surf. Sci. 439 (2018) 468–474, https://doi.org/10.1016/j.apsusc.2018.01.034.
40. A. Saboori, A. Aversa, F. Bosio, E. Bassini, E. Librera, M.D. Chirico, S. Biamino, D. Ugues, P. Fino, M. Lombardi, An investigation on the effect of powder recycling on the microstructure and mechanical properties of AISI 316L produced by Directed Energy Deposition, Mater. Sci. Eng. A. 766 (2019) 138360, https://doi.org/10.1016/j.msea.2019.138360.

41. Z. Sun, X. Tan, S.B. Tor, W.Y. Yeong, Selective laser melting of stainless steel 316L with low porosity and high build rates, Mater. Des. 104 (2016) 197–204, https://doi.org/10.1016/j.matdes.2016.05.035.
42. A. Mohammadia, N.A. Enikeev, M.Y. Murashkin, M. Arita, K. Edalati, Examination of inverse Hall-Petch relation in nanostructured aluminum alloys by ultra-severe plastic deformation, J. Mater. Sci. Technol. 91 (2021) 78-89, https://doi.org/10.1016/j.jmst.2021.01.096.
43. J. Mola, E.J. Seo, L. Cho, Correlation between mechanical stability and hardness of austenite in martensite/austenite mixtures, Mater. Sci. Eng. A. 822 (2021) 141687, https://doi.org/10.1016/j.msea.2021.141687.

Chapter 5
Effect of the Melt Pool In-Situ Control on the Microstructure of Laser Directed Deposition 316L Stainless Steel Thin-Walled Parts

5.1 Introduction

With the advancements of modern manufacturing, heavy equipment has grown rapidly and significantly in size, performance, and integration, thereby increasing the demand for manufacturing technology [1–6]. Large components are vulnerable to failures under harsh working conditions [7–9]. This results in the scrapping of entire components, failure, increased production cost, and material waste. Traditional metal forming techniques include casting [10, 11], forging [12, 13], welding [14, 15], and milling [16, 17]. However, problems such as low processing efficiency [18, 19], multiple assembly procedures [20, 21], and high production costs persist [22, 23]. In recent years, additive manufacturing (AM) technology has developed rapidly, with the advantages of layer-by-layer accumulation of materials and rapid formation of arbitrarily complex parts [24–27]. Laser directed energy deposition (LDED) is an AM process in which a focused laser is used to fuse materials by melting as they are deposited [28–31]. Specifically, LDED is performed by predefining the structure of a part using slicing software, melting the metal powder into a melt pool using a high-energy laser, and producing the part layer-by-layer.

In LDED high-energy laser irradiation melts the powder on the substrate, forming a melt pool. The melt pool is characterized by a high temperature, high brightness, small size, and rapid change; its size is a direct reflection of the laser energy density. Under a continuous deposition strategy, numerous molten beads produce a thermal accumulation effect [32]. To address this problem, an online molten-pool monitoring system is required. Zhang et al. [33] used a coaxial high-speed single-camera-based two-wavelength-imaging pyrometry system to monitor the rapidly changing melt pool state for the acquisition and mapping of spatial information of the melt pool. Gaikwad et al. [34] used two coaxial high-speed video cameras and a temperature-field imaging system to monitor the melt pool state (including temperature, shape and size, and splash intensity) and used it as a characteristic parameter

K. Luo et al., *Melt Pool In-Situ Controlling of Laser Additive Manufacturing Technology*, https://doi.org/10.1007/978-981-95-4597-1_5

for training the pore formation model. To determine the mechanisms influencing the instability of the melt pool state, Guo et al. [35] proposed the use of in-situ high-speed X-ray imaging to monitor the melt pool flow and pore formation. Current research mainly focused on monitoring the melt pool state and information acquisition; the influence of the thermal accumulation effect on the melt pool width and influence law on the microstructure of thin-walled components has not yet been clarified.

In addition, challenges remain in the accuracy and processing rate of melt pool control systems. Despite the advances in the monitoring and online control of melt pools, different challenges remain. An S-shaped scanning-path approach (alternating layer-by-layer deposition along the scanning direction) is typically used to deposit thin-walled parts. However, as the robot arm moves, changes in the travelling path change the direction of the melt pool in the camera lens. Owing to the constant change in the direction of the melt pool, the conventional melt pool width extraction method (using a square bounding box at the periphery of the melt pool to determine the width of the melt pool) is not applicable because it is extremely directional. Secondly, in the in-situ melt pool control strategy, the input laser power changes quickly, resulting in large fluctuations in the melt pool width value. Conventional proportional-integral-derivative (PID) algorithms for solving this type of problem suffer from component flatness problems caused by excessive integral overshoot. Thirdly, interlayer cooling time has been proposed to solve the heat build-up problem, which requires the laser to be switched off briefly at the end of each layer deposition. The instantaneous width of the melt pool monitored by the in-situ control system is zero, which causes the PID controller to dramatically increase the laser power at the initial stage of the next layer, leading to integral saturation. This not only causes a sharp oscillation in the melt pool width profile but also influences the control accuracy and efficiency.

To solve the above difficulties, an in-situ melt pool control technique was proposed to assist the LDED of AISI 316 L SS thin-walled parts. In terms of system construction, a CCD camera was used to monitor the melt pool and extract the width based on the maximum tangent circle fit to the image of the melt pool. This effectively avoided the influence of path planning on the moving direction of the melt pool and can function with a variety of deposition paths. Moreover, to avoid the light-off-caused integral saturation causing drastic oscillation to the melt pool curve, a thin-walled coupon was fabricated by continuous laser scanning. It is worth noting that to ensure high accuracy of the in-situ melt pool control, the uniformity of the width was ensured by the adaptive integral-separated PID control (see the appendix for the specific improvement strategies and algorithms), which also solved the problem of excessive integral overshoot and was conducive to the improvement of the flatness of the thin-walled parts.

5.2 Experiment and Characterizations

5.2.1 LDED Process

As a high-quality carbon structural steel, 45 steel is important for fabricating large engineering components. 316L alloys have the advantages of high hardness, wear resistance, and toughness; thus, they are widely used to prepare coatings to improve the service life of components [36]. In this study, 45 steel and 316L alloys (50–100 μm) were used as the substrate and deposited layers, respectively.

Figure 5.1 displays a diagram of the LDED system with closed-loop control; this consists of a fiber laser, KUKA robotic arm, powder feeder, computerized numerical control (CNC) equipment, argon gas cylinder, imaging systems (for melt pool monitoring), and infrared thermal imager. An external computer was used to display the melt pool images and width curves. The powder was melted as a heat source using a 12-kW fiber laser with a wavelength of 900–1200 nm. Based on the CNC equipment program, the KUKA robot arm acted as a motion system to control the movement of the laser processing head. Argon was used as the carrier and protective gas. The laser beam had a spot diameter of 3 mm below the focal point. In addition, as an important acquisition device for the melt cell, the laser included a coaxially mounted Daheng MER2-041-302GM-P camera on the laser processing head,

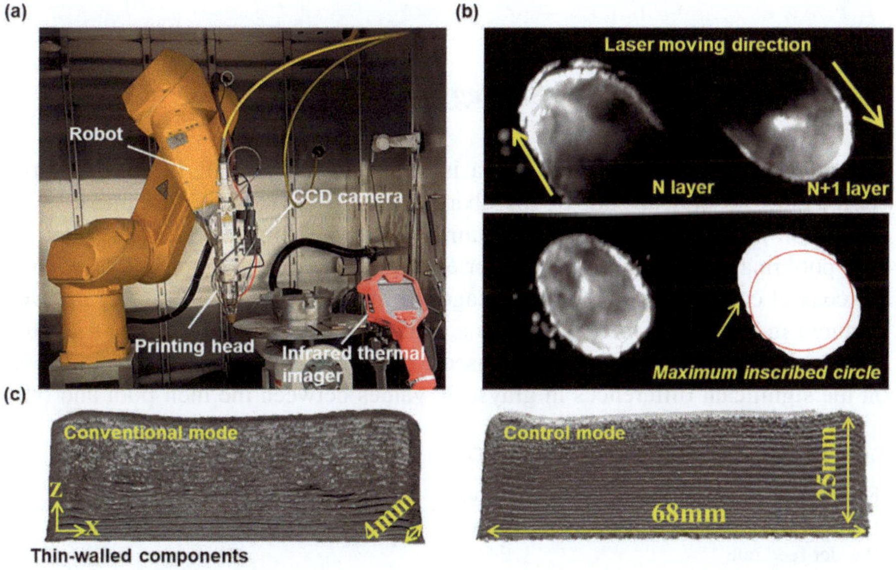

Fig. 5.1 Thin-walled parts prepared by in-situ melt pool control-assisted LDED technology. (**a**) Physical drawings of the system consisting of robotic arm, CCD camera, printing head, infrared thermal imager, (**b**) melt pool images under different scanning paths and the width of the inner tangent circle extracted, (**c**) thin-walled parts with dimensions of 68 × 25 × 4 mm^3 prepared in two modes

coaxial imaging lens, 45° reflector, half-transverse beam splitter, 45° half-transmissive LED light source, and coaxial LED light source with a collimator. To obtain an appropriate brightness and clear image, aperture and focal length adjustments were mounted on the coaxial imaging lens to adjust the amount of light passing through the lens. The collimator allowed more light to pass through the optical path of the laser head and thus be emitted from the small hole in the nozzle and returned. The CCD camera captured images at 50 fps with a resolution of 720 × 540 pixels. An infrared thermal imager placed on the side was used to monitor and collect the real-time temperatures of the components for subsequent temperature gradient and cooling rate calculations. The infrared thermal imager (Fotric 340+) was equipped with a 13-megapixel industrial camera with a resolution of 640*480, a temperature measurement accuracy of ±2 °C, and an imaging distance range of 0.25–50 m. The MagicThermal visible light thermometry module was used, which allowed the detection area to be presented as a color image and other areas to appear as monochrome images. According to the specific deposited material, the emissivity can be set between 0 and 1. The emissivity was set at 0.6 based on the previous study of deposition temperatures of AISI 316L SS [37]. The deposition process was continued until the entire sample was deposited with no set cooling time between the layers, and the idle time was set as zero when depositing thin-walled parts. The overlapping rate was 50%, and the robotic arm was raised upward by 0.8 mm at the end of depositing each layer to meet the defocusing amount requirement. The process parameters are listed in Table 5.1.

5.2.2 Melt Pool Control Strategy

The mounting position of the camera is indicated in Fig. 5.1. In the deposition equipment, the optical path of the charge-coupled device (CCD) camera in the deposition head was integrated to obtain the top view of the melt pool. First, the melt pool image was calibrated. Two orientations were found in the field of view of the coaxial camera (Fig. 5.1b) and images of the deposited thin-walled parts were captured in cross-scan mode to obtain a complete melt pool image. The triangular threshold segmentation method was used to determine the melt pool region based on the significant differences in grayscale values between the melt pool and other

Table 5.1 Parameters of deposition in LDED

Parameters	Value	Units
Laser power	2000	W
Powder feed rate	24	g/min
Scanning speed	20	mm/s
Idle time	0	s
Height increment	0.8	mm
Overlapping rate	50%	/
Spot diameter	3	mm

regions. Moreover, the contours of the region of interest were smoothed using open operations during the morphological processing, thereby yielding a melt pool width image. The inner-connected circle algorithm was used to extract the melt pool width owing to the random orientation of the melt pool movement. Finally, the pixel values and working distance were used to calculate the melt pool width [38]. During the deposition process, the response speed of the in-situ control system required only milliseconds to complete. The image processing speed after image acquisition was 1000 fps; thus, the entire process required 1 ms. In addition, image acquisition, image processing, and analog signal transmission generated a time lag of approximately 50 ms.

Figure 5.2 displays the flowchart of the in-situ melt pool control-assisted LDED based on the integral separation of the PID algorithm. First, the average width of the first five deposited layers was calculated and set to the setpoint value $w_s(t)$. Starting with the sixth layer, the width acquired by the CCD camera was set to a real-time value $w_r(t)$. The difference $e(t)$ between $w_s(t)$ and $w_r(t)$ was used as an input to the entire closed-loop control system. A stable melt pool width was obtained by varying the laser power; the system was discretized to set the variation in laser power to

$$\Delta P = K_P \left\{ e(k) + \beta \frac{1}{T_I} \sum_{k=1}^{n} e(k) + T_D \left[e(k) - e(k-1) \right] \right\}, \tag{5.1}$$

where the PID parameters (K_P, K_I, and K_D) determine the sensitivity of the feedback control system. The range of values of β satisfies

$$\beta = \begin{Bmatrix} 0, |e(k)| > X \\ 1, |e(k)| \le X \end{Bmatrix}, \tag{5.2}$$

where X is the threshold $e(k)$. When $e(k)$ exceeded the threshold range, the separate-integral control mode was used, and the PD (without integral) control was only retained to avoid excessive overshooting. When $e(k)$ was within the threshold range, PID control was used to ensure the stability of the melt pool width curve. The integral separation PID control can avoid the phenomenon of sudden increase in the

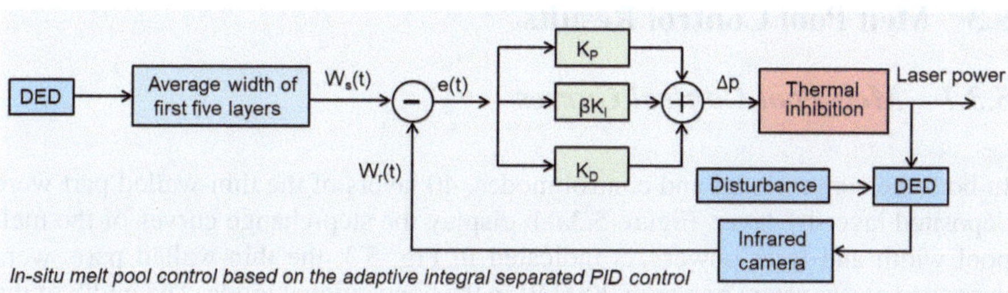

Fig. 5.2 The PID adaptive algorithm for realizing the in-situ melt pool control, which includes the reference width setting, PID parameters, laser power output, the real-time melt pool width of the feedback and the disturbance of the experimental environment

edge of the component produced by integral overshooting, which improves both the response speed and the deposition accuracy. In this study, X was set to 0.5 mm.

The PID parameters were obtained by empirical field tuning. First, K_i and K_d were set to zero, and K_p was adjusted until the system gradually stabilized after the three-step responses were input. K_p was typically set to 0.7 times the K_p that produced the first-step response. Secondly, K_i was adjusted until the oscillation of the system slowed, which was considered the appropriate K_i. Finally, the value of K_d was typically zero or 0.05. In conclusion, the PID parameters must be determined by the fluctuation range of the standard melt pool width; in this case, these were determined to be $K_p = 50$, $T_i = 800$, and $T_d = 0.05$.

5.2.3 Materials Characterization

In this study, thin-walled parts were prepared using LDED in the conventional and control modes. Five thin-walled parts in each of the two modes were fabricated for parallel specimens. The used XYZ coordinate system was shown in Figs. 5.1c and 5.3c, with the length of the specimen shaped along the deposition direction as X, the width as Y, and the height as Z. Several dimensions of 68 mm × 25 mm × 4 mm (length ×height× thickness) thin-walled parts were deposited layer by layer in a "S" scanning path. The samples for microstructure and EBSD characterization were taken from the Y0Z plane, while for surface roughness were taken from the XOZ plane. Cross-sections of all specimens were cut by wire-cut electric discharge machine, and then ground, polished to remove the surface stress. The microstructure of the thin-walled parts and morphology of the fracture surface were photographed using an S-3400N scanning electron microscope (SEM). The extracted samples for EBSD observation were shake-polished in an alumina suspension for 8 h after the standard metallographic polishing process. The grain size and microstructural evolution patterns were obtained using electron backscatter diffraction (EBSD; Gemini SEM 500).

5.3 Melt Pool Control Results

5.3.1 Melt Pool Control Curves

In both the conventional and control modes, 40 layers of the thin-walled part were deposited layer-by-layer. Figure 5.3a, b display the step-change curves of the melt pool width and laser power. As indicated in Fig. 5.3, the thin-walled parts were deposited at a constant power of 2000 W in the conventional mode. The width of the melt pool, as extracted and recorded by the camera, indicated a continuous increasing trend, eventually increasing from approximately 2.3 mm to approximately

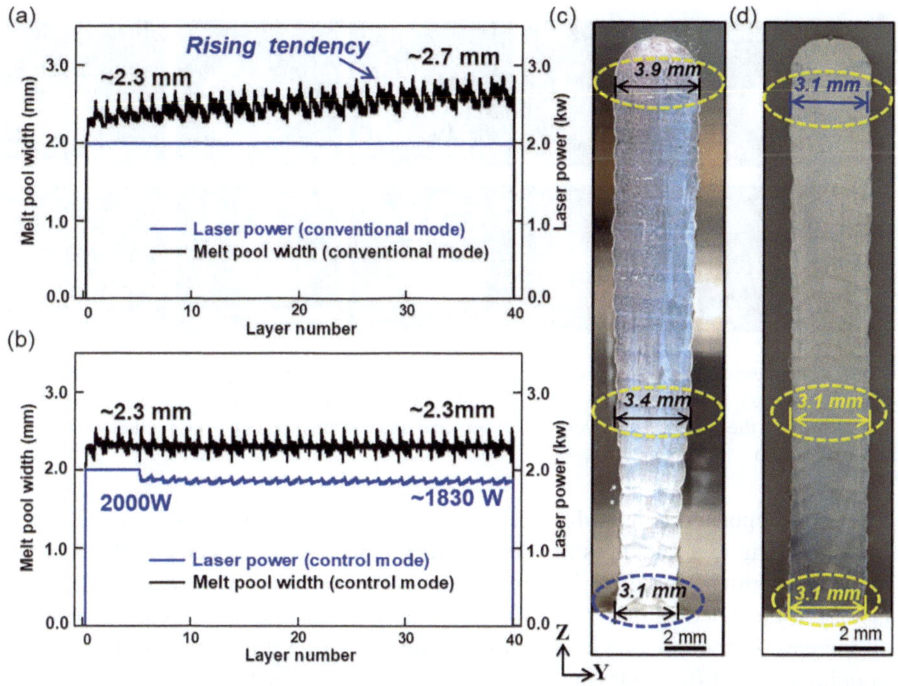

Fig. 5.3 Laser power and melt pool width curves, and cross-section characteristics of thin-walled parts deposited in two modes. (**a**) The melt pool width gradually increases from 2.3 mm to 2.7 mm in the conventional mode, (**b**) the melt pool width remains stable in the control mode, and the laser power decreases in a staged manner to ~1830 W, (**c**) the cross-sectional width of the thin-walled parts increases from 3.1 mm to 3.9 mm in the conventional mode, and (**d**) the cross-sectional width of the thin-walled parts remains at 3.1 mm in the control mode

2.7 mm. Different printing strategies were used in the control mode (Fig. 5.3b). The first five layers were deposited at a constant power of 2000 W, with the average melt pool width at this stage as the set point. The increasing width trend was suppressed by changing the laser power to avoid an excessive increase in the melt pool width. As indicated in the curve, to ensure a stable melt pool width of 2.3 mm, the laser power was reduced from 2000 W to approximately 1830 W.

Figure 5.3c, d displays the cross-sectional profile images in the YZ direction of the thin-walled part. As indicated in Fig. 5.3c, the thickness of the thin-walled part deposited in the conventional mode was not uniform; rather, it increased gradually from the bottom to the top of the part. This is consistent with the trend observed in the width curves. Figure 5.3d displays the cross section of the part in the control mode. The overall uniformity of the part was superior to that in the conventional mode, and its thickness was maintained at approximately 3.2 mm.

This finding was further validated by the changes in the shape of the melt pool indicated in Fig. 5.4, which were obtained during the deposition of different layers (corresponding to the bottom, middle, and top of the component). When the first

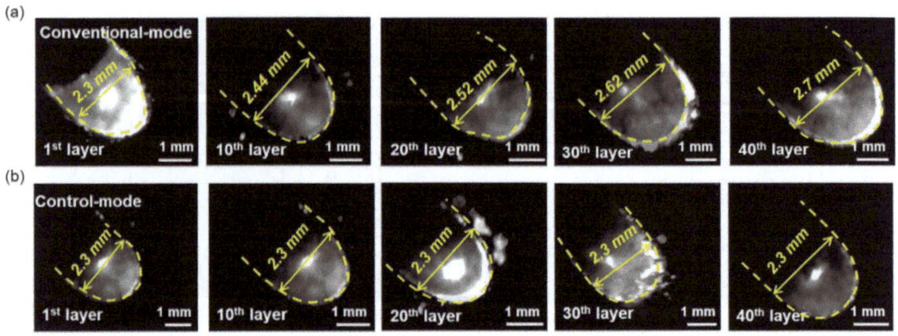

Fig. 5.4 Camera-acquired images of the melt pool during deposition of different layers (collected every ten layers). In conventional mode (**a**), the melt pool width gradually increases from 2.3 mm to 2.7 mm as the number of layers increases, and in control mode (**b**) the melt pool width is always 2.3 mm

layer was deposited (Fig. 5.4a), the width of the extracted melt pool was 2.3 mm, corresponding to a thickness of 3.1 mm. As the deposition proceeded, the detected melt pool width increased to 2.52 mm in the 20th layer. During the deposition of the last layer, the detected melt pool width increased to 2.7 mm, which corresponded to an increase in the thickness of the thin-walled part to 3.9 mm. Similarly, the melt pool images of different layers in the control mode were selected and the melt pool widths were stable at 2.3 mm at different stages and the thickness of the thin-walled parts was held at 3.1 mm at all times. As mentioned by Aggarwal et al. [39] and Sun et al. [40], the consistency of the overall thickness of the thin-walled part contributes to an improvement in the forming quality and overall performance of the deposited part. The PID closed-loop control process proposed in this study helped improve the stability of the melt pool width, thus ensuring the consistency of the overall thickness of the thin-walled part. A careful comparison of the interlayer solidification behavior demonstrated in Fig. 5.1c, d revealed that the interlayer stacking in the control mode was clearer and more readily apparent. This was related to the deposition and solidification behavior of each layer in both modes.

5.3.2 *Melt Pool Control Characteristic*

The width and power variation curves for depositing the 30th–40th layer are displayed in Fig. 5.5a, b, respectively. To illustrate the characteristics of the conventional and control modes, using the 38th layer as an example, the melt pool width versus laser power data for depositing a single layer are displayed separately (Fig. 5.3c–e). As indicated in Fig. 5.5a, b, several prominent peaks could be observed in both modes (marked with dashed lines), which were induced by the deposition direction and integral saturation and could not be eliminated [41]. The thin-walled part was deposited using alternating scans, in which each layer was deposited and

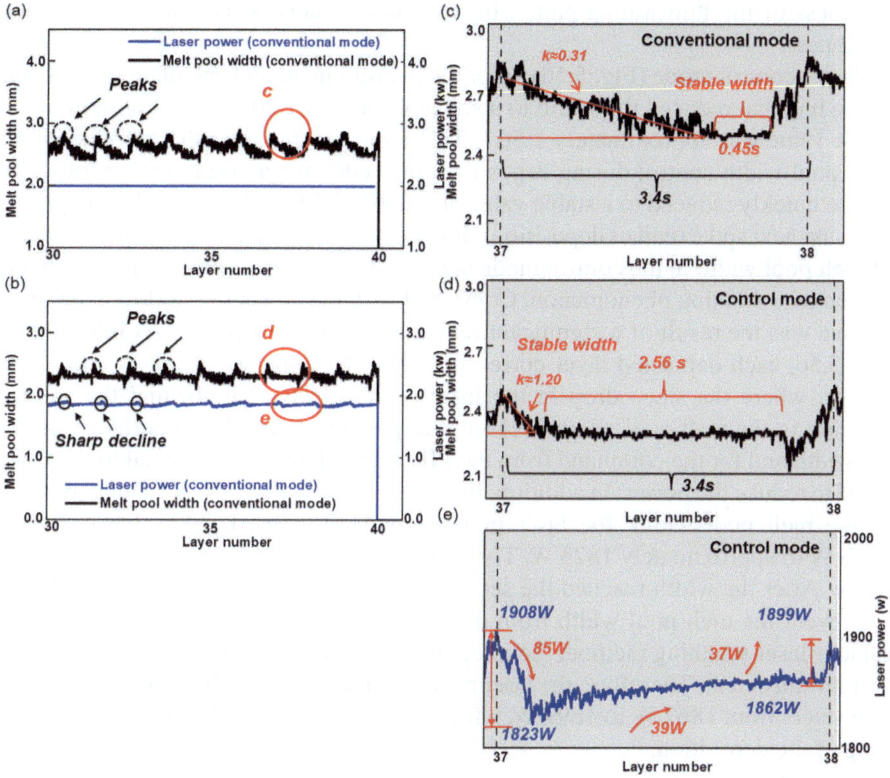

Fig. 5.5 Step curves of melt pool width and laser power for last ten layers deposited in conventional mode (**a**) and control mode (**b**). The appearance time of the stable melt pool width in a single layer in the conventional mode (**c**) is significantly shorter than that in the control mode (**d**) owing to the fact that the laser power was dynamically decreased after the control, which further mitigates the thermal accumulation effect (**e**)

then used as the starting point for the next layer. Thus, the melt pool width at both ends increased abruptly, producing sharp peaks. The width curves indicate that the width changed in the same manner at both ends of the deposition process, i.e., the melt width increased abruptly at both ends. The main difference could be observed in the middle section of each deposition layer. As indicated in Fig. 5.5a, the laser power was constant at 2000 W in the conventional mode. When the scanning speed was constant, the curve indicated that the melt width decreased slowly to a stable value after the spike. As indicated in Fig. 5.5c, the slope of the straight line from the highest to the stable value was approximately 0.31. Based on the length of the deposited layer and scanning speed, the deposition time of each layer was approximately 3.4 s, and the time of the stable melt width was only 0.45 s. The curve appeared to be stable for a short time, and then quickly produced the next spike, which represented the end of the deposition of this layer. This indicates that the width changes rapidly and is not truly stable, leading to nonuniformity in the thickness of the thin-walled part, which further influences the surface finish and roughness.

In the control mode (Fig. 5.5d), after the measurement and calculation, the slope of the line that reduced the width to a stable value was 1.20 and the duration of the stable value was approximately 2.56 s. This indicates that after the introduction of melt pool width control during deposition, the sudden increase in melt pool width can be quickly reduced to a stable value and stabilized for a long time before generating the next spike (end of deposition). It is worth noting that, the suddenly decrease of melt pool width in the control mode at the end of the 2.56 s bracket is to counteract the overshooting phenomenon. Combined with the power curve (Fig. 5.5e), this change was the result of a significant reduction in the laser power. As indicated in Fig. 5.5b, each deposited layer corresponded to the same set of power and width curves, where the steep drop in the laser power always lags behind the sudden increase in the melt pool width because it required time for the camera to monitor the width and for the command from the PID controller to send a signal to the CNC stage to reduce the power. In addition, it can be observed that after a sudden increase of the melt pool width, the laser power rapidly decreased from approximately 1908 W to approximately 1823 W. The width then rapidly decreased to the set value range. After the width reached the set value, the laser power was increased slowly to prevent the melt pool width from decreasing significantly. Because of the continuous laser scanning method, the integral saturation phenomenon was suppressed to a certain extent. Therefore, the laser power increased less at the end of the deposition stage, from 1862 W to 1899 W, which remained less than 1908 W at the beginning of the deposition.

As indicated in Fig. 5.6a, c, the temperature of the melt pool exhibited cyclic variation. The instantaneous temperatures of the melt pool reached approximately 1600 °C in both modes when the deposition time was 50 s. As the number of layers increased, the molten-pool temperature continued to increase in the conventional mode, reaching a maximum temperature of approximately 1923 °C. The lowest temperature was approximately 1542.6 °C at the end of single-layer, and the maximum temperature variation reached approximately 380.4 °C. Conversely, in the control mode, the maximum temperature of the melt pool was detected to be approximately 1739 °C, and the minimum temperature dropped to approximately 1151 °C during the cycle. Compared to the conventional mode, the maximum temperature of the melt pool was significantly lower; the temperature variation reached approximately 588 °C. It can be observed that the overall temperature variation in the controlled mode was larger than that in the conventional mode, which is due to the in-situ control process of the melt pool reducing the laser power input.

Figure 5.6b, d compare the cooling rates of the two modes. In the conventional mode, the highest cooling rate of the melt pool was in the first layer, which reached approximately 3019.5 °C/s because the heat of the first layer was dissipated quickly at the ambient temperature. As the number of layers increased, the overall cooling rate demonstrated a dynamic decrease. The lowest cooling rate of approximately 435 °C/s was observed when the layers were deposited up to approximately 225 s. This coincided with the point at which the highest melt pool temperature was

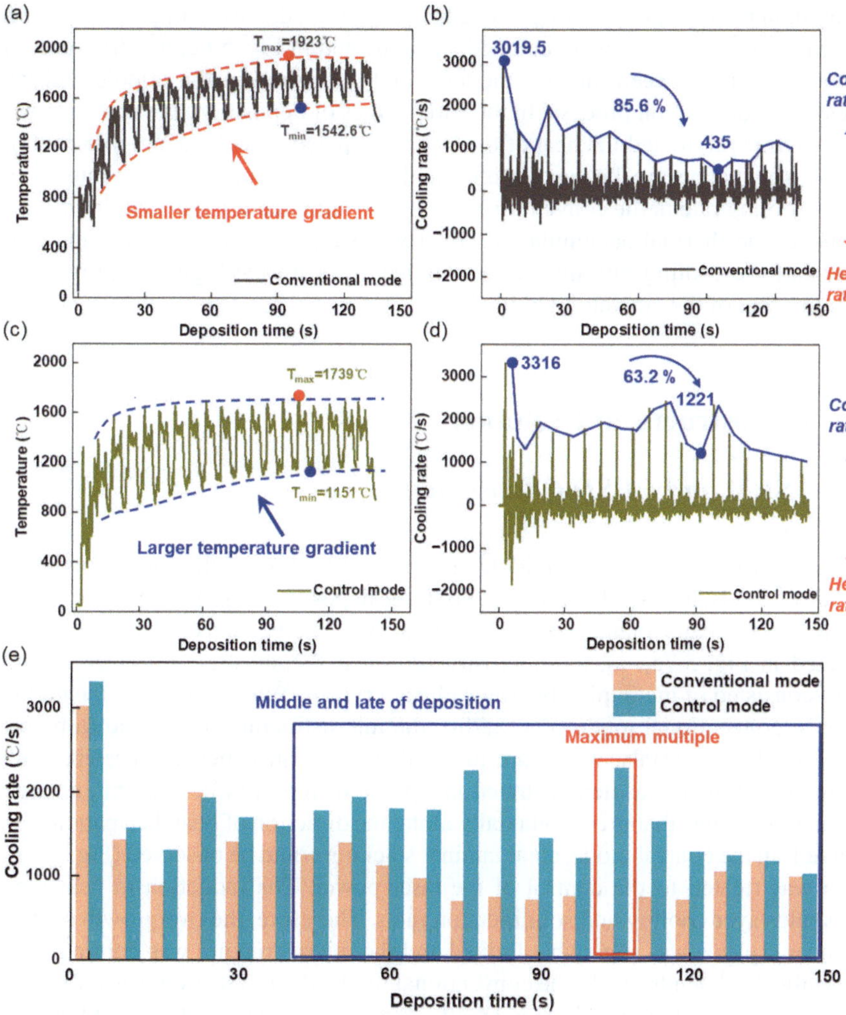

Fig. 5.6 Thermal cycle curves and cooling rate statistics for two modes. (**a**) The thermal cycling temperature of the deposition process in the conventional mode reaches a maximum of 1923 °C, and (**b**) corresponding to a cooling rate decrease of 85.6%. (**c**) The thermal cycling temperature of the deposition process in the control mode reaches a maximum of 1739 °C, and (**d**) the cooling rate in the control mode decreases by 63.2%. (**e**) The comparison of the cooling rates in the two modes shows the effect of the control mode for improving the cooling rate

detected, as indicated in Fig. 5.6a. In the conventional mode, the decrease in cooling rate reached 85.6%. In the control mode, the cooling rate of the first layer reached 3316 °C/s. The overall cooling rate floated near 2000 °C/s as the number of layers increased. Among them, the lowest cooling rate of 1221 °C/s was achieved at the 29th–30th layers, which was an increase of 180% relative to that in the conventional mode.

Based on the real-time cooling rate data in Fig. 5.6b, d, the cooling rates of every two layers in both modes were counted. As indicated in Fig. 5.6e, the difference in the cooling rate between the two modes mainly occurred in the middle and late stages of the deposition process. In the early stages of deposition, the substrate was at room temperature and the heat of the layer dissipated quickly at a higher cooling rate. With an increase in the deposited layer, the heat accumulation effect intensified and the cooling rate in the conventional mode was significantly reduced. In the control mode, the thermal accumulation effect was suppressed by reducing the laser power and the cooling rate increased significantly, with the highest rate reaching five times that of the conventional mode.

5.4 Microstructural Evolution

5.4.1 SEM and EBSD Observation

The microstructure of the top, middle, and bottom section of the specimens in the YOZ direction is displayed in Fig. 5.7. For laser-deposited AISI 316L SS, the microstructure is typically planar crystal-columnar dendrite-equiaxed crystals [42]. As indicated in Fig. 5.7a_1, a_2, the microstructure near the substrate interface in both modes consisted of incompletely grown planar crystals. When observing the middle region in conventional mode (Fig. 5.7b_1), the microstructure was found to be primarily dendritic crystals, which are more likely to form under non-preferentially directed heat transfer, as demonstrated in a previous report [43]. In addition, larger columnar dendrites that grew epitaxially along the direction of heat dissipation were observed. In the control mode, clear laminar stacking could be observed (Fig. 5.7b_2), which was related to the control of the laser power. The reduction in heat input drives interlayer cooling and solidification rates. Therefore, the overgrowth of large columnar crystals was somewhat limited and the microstructure was dominated by uniformly sized dendrites. In the conventional mode, the top section was influenced by thermal accumulation and the microstructure continued to be dominated by dendrites and a small number of equiaxed crystals (Fig. 5.7c_1). The dendrite size was smaller and tended to be nearly equiaxed after the control treatment (Fig. 5.7c_2). The number of small equiaxed crystals increased significantly, indicating clear grain refinement.

Based on the microstructure and melt pool width curves, it was found that the main difference between the two modes was reflected in the mid-top region. To investigate the reason for the different thicknesses in the two modes, Fig. 5.8 displays the EBSD results for selected samples from the middle section in both the conventional and control modes. As can be observed from Fig. 5.8a, b, when depositing 40 layers, a clear solidification structure was deposited layer-by-layer in the control mode. It can be observed that the melt pool angle was approximately 0° in the parallel state in the conventional mode, whereas the melt pool angle was

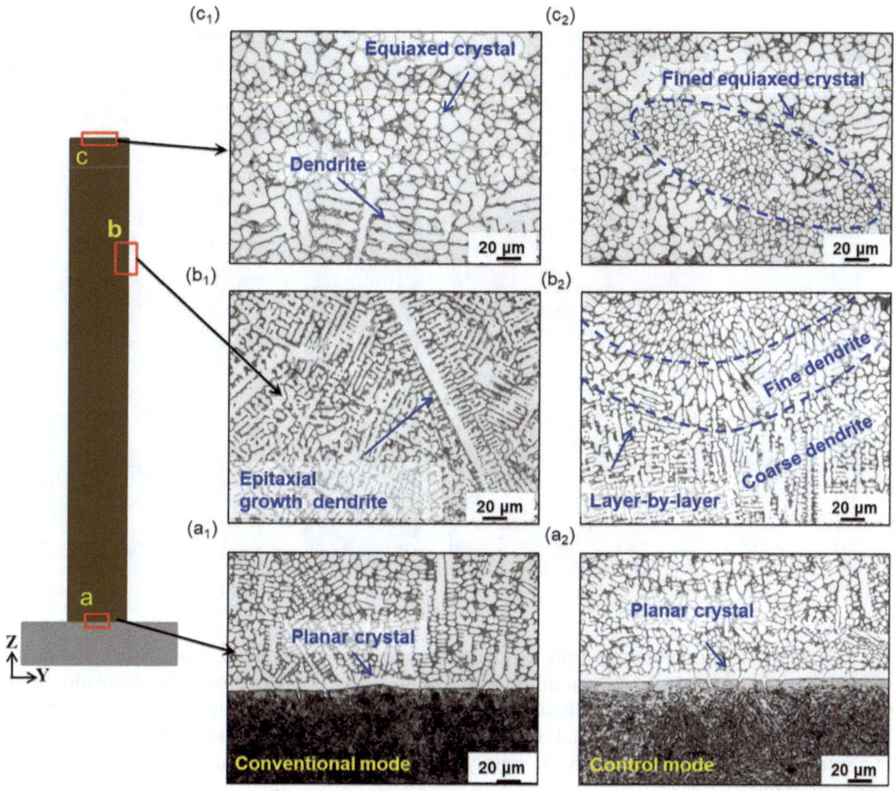

Fig. 5.7 Microstructure of YOZ direction for thin-walled parts at bottom, middle, and top sections in conventional mode (**a**$_1$, **b**$_1$, **c**$_1$) and control mode (**a**$_2$, **b**$_2$, **c**$_2$). The microstructures at the bottom of both modes (a_1, a_2) are dominated primarily by planar crystals. The specimens in the middle (**b**$_1$) and top (**c**$_1$) of the conventional mode are columnar dendrites and equiaxed grains, respectively, whereas the dimensions of grains in the middle (**b**$_2$) and top (**c**$_2$) section of specimens in the control mode are significantly refined

approximately 16° in the control mode. This is related to the different temperature gradients in the two modes, which influence the direction of the Magorani flow. When the same size was selected for EBSD analysis (labeled as a_1, b_1), only three layers of the solidified structure were observed in the conventional mode, whereas five layers were observed in the control mode. Because the lengths of the interlayer dendrites were influenced by the cooling and solidification rates. As indicated in Fig. 5.8a_1, the continuous evolution of dendrite-equiaxed crystals was demonstrated in the interlayers. The growth region of the interlayer equiaxial crystals was short and covered by large dendrites. The three dendrite sizes were approximately 463 μm, 621 μm, and 642 μm. Figure 5.8b_1 displays the continuous evolution of interlayer dendritic and equiaxed crystals in the control mode, where there is a clear reduction in the length of the interlayer dendritic crystals compared with the conventional mode. The continuous equiaxed crystals truncated the large epitaxially

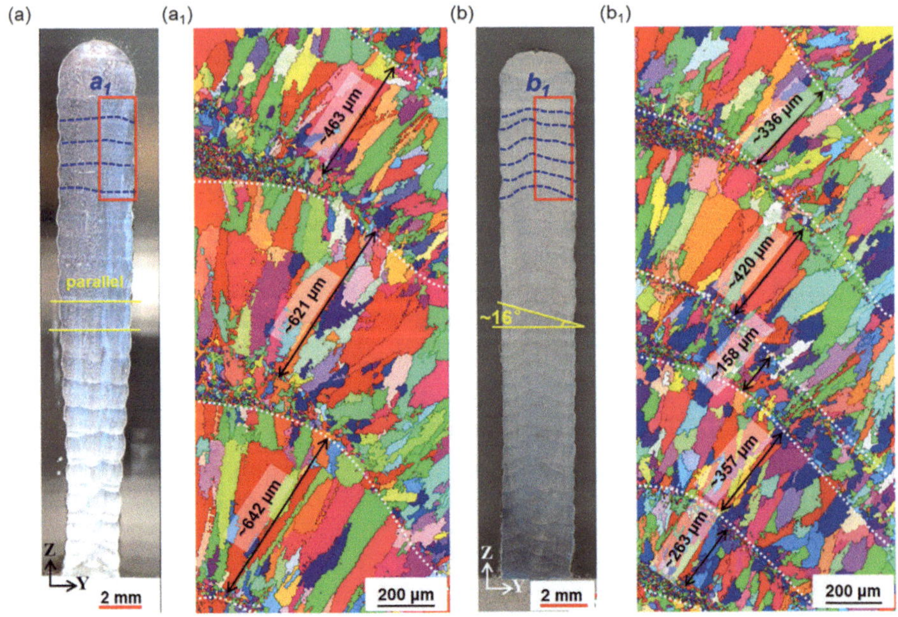

Fig. 5.8 Cross-sectional morphology and EBSD results of specimens in both modes. The cross-sectional width (**a**) of the specimen in the conventional mode increases with the deposition time, whereas the interlayer ($\mathbf{a}_1$) is distributed mainly with large-sized dendrites and equiaxed grains. The uniformity of the cross-sectional width (**b**) of the specimen in the control mode is significantly increased, whereas the interlayer ($\mathbf{b}_1$) is mainly a periodic distribution of small-sized dendrites and equiaxed grains

grown dendrites, thus exhibiting a clear laminar structure. The lengths of interlayer columnar dendrites were typically 1/3 to 1/2 of those in the conventional mode, mainly approximately 263 μm, 357 μm, 158 μm, 420 μm, and 336 μm.

5.4.2 *Effects Factors of Microstructural Evolution*

The in-situ melt pool control process can influence the development of the microstructure by increasing the cooling rate, thereby enhancing the dimensional accuracy of thin-walled parts. Taking the YOZ cross section of the LDEDed thin-walled part as an example, Fig. 5.9 displays the microstructural distribution and characteristics of AISI 316L SS in both modes. Except for the planar crystals produced at the bond with the substrate, the microstructure indicates the continuous evolution from columnar dendrites to equiaxed dendrites [44]. At the initial stage of deposition, when the cooling rate was sufficiently high and the heat dissipation was fast, there was no obvious difference in the distribution and width of the bottom microstructure of the samples in two modes. Although the fixed height-off of the laser head was set in two modes, there was a significant difference in the layer thickness between the middle and top of the specimens. This is related to the different cooling rates in two modes. Depending on the temperature gradient (*G*) and growth rate (*R*) of the melt pool during deposition (Fig. 5.9b), there were two combinations: *GR* and *G/R* (cooling rate) [45, 46]. The cooling rate was significantly reduced in the middle of the conventional mode. Because the deposition was continuous (with no interlayer cooling time), the deposition of the (n + 1)th layer started before the *n*th layer was completely cooled and solidified. This resulted in the growth of columnar crystals through the region of virtually two-layer thickness before reaching the solidification rate and transforming into equiaxed grains. Therefore, although the number of deposited layers was 40, no clear solidification traces were observed for each layer in the cross section of the specimen (Fig. 5.8a).

In addition, owing to the variation in the cooling rate influencing the growth of columnar dendrites, the widths increased in the cross sections of the thin-walled parts. In the conventional mode, the cooling rate in the middle and late stages of deposition decreased up to 85.6% compared with the early stages. A sharp decrease in the cooling rate promotes excessive epitaxial growth of the interlayer dendrites, thus limiting the growth of equiaxed crystals. Owing to the slow cooling rate, the

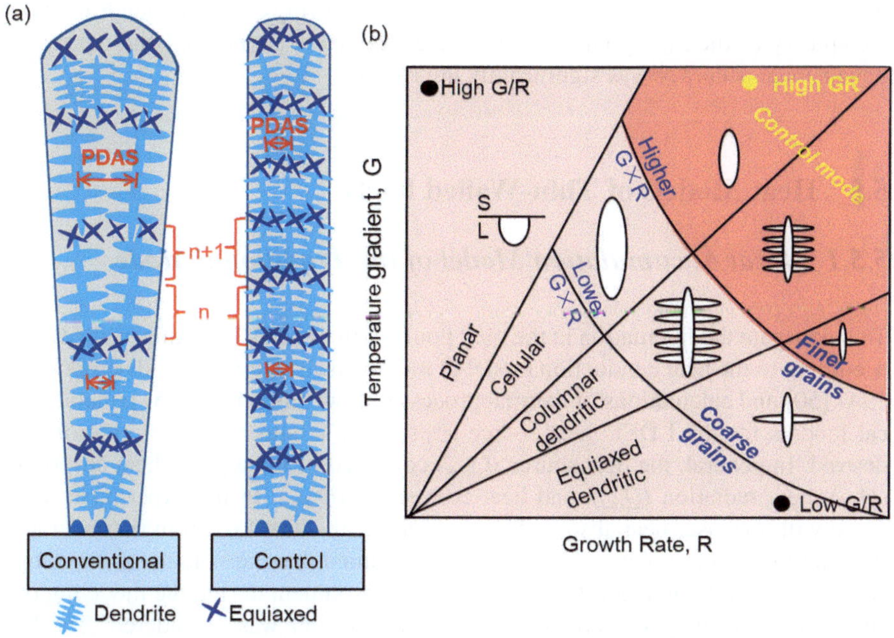

Fig. 5.9 Schematic diagrams of distribution and dimensional changes of microstructure for thin-walled parts in two modes. (**a**) Characteristics of microstructure distribution of YOZ cross-section: interlayer columnar dendrites with large distribution area, large layer thickness, and sparse equiaxed crystals in conventional mode; alternately distributed interlayer columnar crystals and equiaxed crystals, small layer thickness in the control mode. (**b**) Correspondence between temperature gradient, cooling rate, and microstructure

microstructure did not reach the solidification temperature before the next layer was deposited, and the dendrites grew in the direction of heat dissipation from the melt pool. Therefore, the length of the epitaxial growth dendrites increased constantly. Based on the common empirical equation between the cooling rate of the iron-based alloy ($\dot{T}$) and primary dendrite arm spacing (*PDAS*) [47], $PDAS = 100\dot{T}^{-0.35}$, where −0.35 is a constant. A decrease in the cooling rate led to an increase in the *PDAS*, and a gradual increase in the thickness of the thin-walled member was observed in the YOZ direction.

In the control mode, the cooling rate was significantly increased by dynamically decreasing the laser power; the minimum cooling rate was 2.8 times higher than that in the conventional mode. This not only accelerates melt-pool solidification but also promotes the rapid transformation of monolayer columnar grains to equiaxed grains. The ratio of the temperature gradient to the solidification rate, *G/R*, is the key factor that determines the solidification microstructural morphology (planar, cellular, columnar, or equiaxed grains) [48, 49]. As the cooling rate increased, the transition from columnar to equiaxed grains accelerated, and the interlayer exhibited a regular alternating distribution of columnar dendrites and equiaxed grains. By analyzing the cross-sectional profile (Fig. 5.8b), clear solidification could be observed in the 40 deposited layers of the specimen. In addition, the increase in cooling rate also contributed to the further reduction of the PDAS; the growing columnar dendrites along the epitaxy of the melt pool boundary were effectively suppressed and the overall dimensional accuracy was significantly improved.

5.5 Heat Models of Thin-Walled Parts

5.5.1 Heat Accumulation Model of the Traditional Mode

To investigate the fluctuation in the melt pool width of AISI 316L thin-walled parts, according to the heat conduction model of moving heat source in solids proposed in 1947 [50] and calculations of thermal processes in welding theory [51], mathematical models for the LDED heat source assisted by in-situ melt pool control were derived. In general, the heat source (Q_{heat}) consisted of three parts: heat absorption (Q_{abs}), heat radiation (Q_{rad}), and heat convection (Q_{conv}). In the experiment, compared with the laser heat absorbed by the sample, the heat loss from thermal radiation and thermal convection was small. We assumed that the volume of the sample was initially in thermal equilibrium with the environment during the modeling process; therefore, only the effect of thermal absorption was considered. The heat absorbed by a single layer

$$Q_{heat} = \eta_{abs} \cdot P \cdot t, \tag{5.3}$$

which is related to the laser power (*P*), heat absorptivity (η_{abs}) and time (*t*). The scanning speed *v* was set to be constant in advance of the experiment such that the time for each deposition layer could be calculated using $t = \frac{l}{v}$, where l is the length of the deposition layer. The absorbed heat was further transformed into an equation related to the scanning speed as follows:

$$Q_{heat} = \frac{\eta_{abs} \cdot P \cdot l}{v}. \tag{5.4}$$

According to the different shapes of the heat-source models proposed by Shang et al. [52], the geometry of thin-walled parts applied to a one-dimensional heat flow from the surface heat source of a semi-infinite object. The area of each pass of the melt pool was typically set to $A = l \cdot w$, and the surface heat source was normalized to the geometry.

$$Q_{1D} = \frac{Q_{heat}}{A} = \frac{\eta_{abs} \cdot P}{v \cdot w} \tag{5.5}$$

is quantified in terms of the energy-to-area ratio A. The heat input to LDED was continuous during deposition, and each superposition of the layers could be considered as the frequency $f = \frac{1}{t} = \frac{v}{l}$ of the energy input during the construction of the component. The frequency of the input (i.e., the number of layers) is related to the heat accumulation effect; thus, the normalized energy can be further transformed into

$$Q_{1D} = \frac{\eta_{abs} \cdot P}{f \cdot l \cdot w}. \tag{5.6}$$

Each deposition layer consisted of several individual beads; the input frequency of the bead

$$f_{bead} - \frac{1}{t} = \frac{1}{N_{bead} \cdot \frac{d}{v}} \tag{5.7}$$

is related to the number of melting pools (N_{bead}), the length of the individual pools (d), and the scanning speed (v). Then, Eq. (5.7) is inserted into Eq. (5.6) to obtain the heat absorbed by a single melt pool after normalization

$$Q_{bead} = \frac{\eta_{abs} \cdot P}{f \cdot N_{bead} \cdot d \cdot w}. \tag{5.8}$$

Based on the thermal accumulation model proposed by Hagenlocher et al. [53], we improved and presented the local temperature increase in a single melt pool using

$$\Delta T(r,t) = \frac{Q_{bead}}{\rho \cdot c_p \cdot \sqrt{(4\pi k t)}} \cdot e^{-\frac{r^2}{4k} \cdot \frac{1}{t}}, \tag{5.9}$$

which changed by the deposition time (t) and scanning position (r). ρ is density, c_p is the heat capacity, and $k = \dfrac{\lambda}{\rho \cdot c_p}$ is the heat diffusivity. λ is the thermal conductivity of the material. Owing to the discrete character of the repetitive input frequency in each layer or independent bead, the heavy-side function

$$\theta\left(t-\frac{N-1}{f}\right)=\begin{cases}0, t-\dfrac{N-1}{f}<0, N<1\\ 1, t-\dfrac{N-1}{f}\geq 0, N\geq 1\end{cases} \tag{5.10}$$

is used, the values of which are related to N. When $t-\dfrac{N-1}{f}<0$, the function takes the value of zero; when $t-\dfrac{N-1}{f}\geq 0$, the function takes the value of one. The temperature increase

$$\Delta T(N,r)=\sum_{N=1}^{N_A}\Delta T\left(r,t-\frac{N-1}{f}\right)=\frac{Q_{\text{bead}}}{\rho\cdot c_p\cdot\sqrt{4\pi k}}\cdot\sum_{N=1}^{N_A}\frac{\theta\left(t-\frac{N-1}{f}\right)}{\sqrt{t-\frac{N-1}{f}}}\cdot e^{\frac{-r^2}{4k}\cdot\frac{1}{t-\frac{N-1}{f}}} \tag{5.11}$$

of all melt pools in each layer with increasing time and position is correlated with the frequency f, the number of energy inputs N, and the total number of energy inputs N_A. During the actual deposition process, it was observed that a minimum temperature existed at the end of each layer in the thermal cycling profile (Fig. 5.5a, c) and that the minimum temperature increased dynamically for each layer. This is defined as the residual temperature that exists briefly prior to the next energy input. The residual temperature influences the cooling and solidification states of the deposited layers. The difference between the residual temperature of the previous layer and that of the next layer is defined as the residual temperature difference ΔT_R. As proposed by Weber et al. [54], the transient presence of the residual temperature is expressed in terms of an infinitesimally short period of time $\dfrac{\delta}{f}$, and therefore the residual temperature exists for a time $t=\dfrac{N_A-\delta}{f}$, which can be substituted into Eq. (5.10) as

$$\Delta T_R(N,r)=\sum_{N=1}^{N_A}\Delta T\left(r,t-\frac{N-\delta}{f}\right)=\frac{Q_{bead}}{\rho\cdot c_p\cdot\sqrt{4\pi k}}\cdot\sum_{N=1}^{N_A}\frac{1}{\sqrt{\frac{N-\delta}{f}}}\cdot e^{\frac{-r^2}{4k}\cdot\frac{1}{\frac{N-\delta}{f}}}. \tag{5.12}$$

As $\delta \to 0$, Eq. (5.12) further simplifies to

$$\Delta T_R\left(N_A, r\right)=\frac{Q_{bead}}{\rho \cdot c_p \cdot \sqrt{4\pi k}} \cdot \sum_{N=1}^{N_A} \frac{1}{\sqrt{\frac{N}{f}}} \cdot e^{\frac{-r^2}{4k} \cdot \frac{1}{\frac{N}{f}}}. \tag{5.13}$$

During the deposition of the thin-walled parts, the heat source moves with the coordinates. In the r-coordinate system, the sum of the residual temperature distributions of the entire component is

$$\begin{aligned} \Delta T_R\left(N, r\right) &= \sum_{N=1}^{N_A} \Delta T\left(r_N = r - Ns, t = \frac{N}{f}\right) \\ &= \frac{Q_{bead}}{\rho \cdot c_p \cdot \sqrt{4\pi k}} \cdot \sum_{N=1}^{N_A} \frac{1}{\sqrt{\frac{N}{f}}} \cdot e^{\frac{-(r-Ns)^2}{4k} \cdot \frac{1}{\frac{N}{f}}}, \end{aligned} \tag{5.14}$$

which represents the residual temperature distribution between different layers. The initial ambient temperature T_0 and the sum of all the residual temperature differences constitute the residual temperature of the thin-walled part as follows:

$$T_R\left(N, r\right) = T_0 + \frac{Q_{bead}}{\rho \cdot c_p \cdot \sqrt{4\pi k}} \cdot \sum_{N=1}^{N_A} \frac{1}{\sqrt{\frac{N}{f}}} \cdot e^{\frac{-(r-Ns)^2}{4k} \cdot \frac{1}{\frac{N}{f}}} \tag{5.15}$$

which is the cumulative thermal effect of the entire component as a function of power, melt pool width, position, and time. The in situ control of the melt pool width is a dynamic correction process over time. To analyze the effect of heat accumulation per unit volume of the component per unit time, the volume difference of the deposited independent beads per unit time is defined as

$$\Delta \dot{V} = \frac{P \cdot \eta_{abs}}{\rho \cdot \left(\Delta T_R \cdot \mathrm{c_p} + h_s\right)} \tag{5.16}$$

For a given volume of a single melt pool with thickness s, length d, and width w, the maximum possible frequency

$$f = \frac{\Delta \dot{V}}{s \cdot d \cdot w} \tag{5.17}$$

is defined as the difference in volume per unit of time. By substituting Eqs. (5.8), (5.14), and (5.17) into Eq. (5.16) and defining $f(N)=\sum_{N=1}^{N_A}\frac{1}{\sqrt{\frac{N}{f}}}\cdot e^{\frac{-(r-Ns)^2}{4k}\cdot\frac{1}{\frac{N}{f}}}$, we can obtain the relationship between the heat absorbed by a single melt pool per unit time and the width of the melt pool as follows:

$$f\cdot s\cdot d\cdot w=\frac{P\cdot\eta_{abs}}{\rho\cdot\left(\frac{\frac{\eta_{abs}\cdot P}{f\cdot N_{bead}\cdot d\cdot w}}{\rho\cdot c_p\cdot\sqrt{4\pi k}}\cdot f(N)\cdot \mathrm{c_p}+h_s\right)},\tag{5.18}$$

which further simplifies to the equation for laser power versus melt pool width

$$P=\frac{h_s\cdot f\cdot s\cdot\rho\cdot d}{\eta_{abs}\left(1-\frac{s\cdot f(N)}{N_{bead}\cdot\sqrt{4\pi k}}\right)}\cdot w.\tag{5.19}$$

5.5.2 *Heat Accumulation Model of the Control Mode*

Based on the relationship between the laser power and melt pool width in the conventional mode, when the power is held constant, the melt pool width is continuously increased by the residual temperature and thermal accumulation effects. Therefore, to achieve the goal of in situ control of the melt pool width, the laser power must be dynamically adjusted based on the variation in the melt pool width. In industrial applications, PID controllers can control the output by calculating the magnitude and rate of change of the input error, and achieve precise control of the controlled object based on proportionality, integration, and differentiation, which are suitable for systems with a change in dynamic characteristics. For each bead, different laser power inputs result in different widths. During the deposition process, assuming that the number of melt pools is n, the difference between the width of each pool captured in real time and the ideal value provides a reference for the laser power variable of the next pool, as indicated in Eq. (5.20)

$$\left\{\begin{array}{l} 1^{bead}: P_1 \sim w_1(k) \\ e_1(k) = \left|w_s(k) - w_1(k)\right| \\ \Delta P_1 = K_P\left\{e_1(k) + \frac{1}{T_I}\sum_{m=1}^{k} e_1(m) + T_D\left[e_1(k) - e_1(k-1)\right]\right\} \\ P_2 = P_1 + \Delta P_1 \\ \ldots\ldots \\ n^{bead}: P_n \sim w_n(k) \\ \Delta P_n = K_P\left\{e_n(k) + \frac{1}{T_I}\sum_{m=1}^{k} e_n(m) + T_D\left[e_n(k) - e_n(k-1)\right]\right\} \\ P_n = P_{n-1} + \Delta P_{n-1} = P_1 + \sum_{i=1}^{n-1} \Delta P_i \end{array}\right\}. \tag{5.20}$$

In the control-mode deposition stage, 1^{bead} is the number of the first melt pool, P_1 is the initial power, $w_1(k)$ is the actual width of the molten pool under irradiation of 2400 W. $w_s(k)$ is the width of the reference melt pool, $e_1(k)$ is the difference between the actual width of the first molten pool and $w_s(k)$, and is the input to the PID controller. ΔP_1 is the amount of change in the laser power corrected based on the width of the first melt pool and width of the reference value, which is also the output of the PID controller. P_2 is the input of the second laser power in the control mode (the derivation here includes the integral action). When the nth melt pool is deposited, the corresponding parameter settings are similar to those for the first layer. The entire in-situ melt pool control is a process of continuously correcting the laser power with the reference width of the melt pool as the control target [55]. The sum of the laser power inputs after the deposition of the entire component

$$\begin{aligned} P_{all} = \sum_{i=1}^{n} P_i &= nP_1 + (n-1)\Delta P_1 + (n-2)\Delta P_2 + \ldots + \left[n - (n-1)\right]\Delta P_{n-1} \\ &= nP_1 + \sum_{m-1}^{n-1} (n-m)\Delta P_i \end{aligned} \tag{5.21}$$

is the summation equation for the number of pools and laser power corrections. To obtain the laser power per unit time volume of a single melt pool in the control mode, the total power during the deposition of the components is normalized as

$$\dot{P} = \frac{P_{all}}{d \cdot s \cdot w} = \frac{nP_1 + \sum_{m-1}^{n-1} (n-m)\Delta P_i}{d \cdot s \cdot w}. \tag{5.22}$$

Substituting Eq. (5.22) into Eqs. (5.8) and (5.14), the heat absorbed by each melt pool in the control mode and the residual temperature difference are

$$Q_{bead}' = \frac{\eta_{abs} \cdot \dot{P}}{f \cdot N_{bead} \cdot d \cdot w}, \tag{5.23}$$

$$\begin{aligned}\Delta T_R(N,r)' &= \sum_{N=1}^{N_A} \Delta T\left(r_N = r - Ns, t = \frac{N}{f}\right)' \\ &= \frac{Q_{bead}'}{\rho \cdot c_p \cdot \sqrt{4\pi k}} \cdot \sum_{N=1}^{N_A} \frac{1}{\sqrt{\frac{N}{f}}} \cdot e^{\frac{-(r-Ns)^2}{4k} \cdot \frac{1}{\frac{N}{f}}}.\end{aligned} \tag{5.24}$$

The main difference between the conventional model and control mode is the dynamic adjustment of the laser power. Based on the heat accumulation model derived in the conventional mode, the relationship between the heat absorbed in a single melt pool of a component per unit time and the width of the melt pool in the control mode is as follows:

$$\dot{P} = \frac{h_s \cdot f \cdot s \cdot \rho \cdot d}{\eta_{abs}\left(1 - \frac{s \cdot f(N)}{N_{bead} \cdot \sqrt{4\pi k}}\right)} \cdot w. \tag{5.25}$$

In the control mode, P_I is the initial laser power, which is also the constant power in the conventional mode; thus, the laser power per unit time volume in the control mode can be further expressed as

$$\dot{P} = \frac{nP + \sum_{m-1}^{n-1}(n-m)\Delta P_i}{d \cdot s \cdot w}. \tag{5.26}$$

Substituting Eq. (5.25) into Eq. (5.26) provides the relational equation for the heat effect of a single melt pool and the width in the control mode.

$$\frac{nP + \sum_{m-1}^{n-1}(n-m)\Delta P_i}{d \cdot s \cdot w} = \frac{h_s \cdot f \cdot s \cdot \rho \cdot d}{\eta_{abs}\left(1 - \frac{s \cdot f(N)}{N_{bead} \cdot \sqrt{4\pi k}}\right)} \cdot w \tag{5.27}$$

which can be further simplified as

$$P = \frac{h_s \cdot f \cdot s^2 \cdot \rho \cdot d^2 \cdot w^2}{n \cdot \eta_{abs}\left(1 - \frac{s \cdot f(N)}{N_{bead} \cdot \sqrt{4\pi k}}\right)} - \frac{\sum_{m-1}^{n-1}(n-m)\Delta P_i}{n} \tag{5.28}$$

5.6 Summary

In this chapter, in order to reduce the thermal accumulation effect and improve the forming quality of thin-walled component, an in-situ melt pool control-assisted LDED technique was proposed. AISI 316L SS thin-walled parts were prepared in both conventional mode and traditional mode. The temperature variation and cooling rates were compared to verify the role of in-situ melt pool control on thermal accumulation. This technique provides a new reference for the preparation of other metallic materials and different shapes of components.

1. The designed in-situ melt pool control technique was based on monitoring and extracting the melt pool width by a coaxial CCD camera and setting appropriate PID parameters to regulate the laser power to obtain a stable melt pool width. Both the continuous melt pool curves and the hierarchically extracted images confirm the effectiveness of improving the melt pool width.
2. Compared with the conventional mode, the stability of the melt pool width after control was improved by 17.4%, and the dimensional accuracy of the components was improved by 25.8%.
3. The reasons for the improvement in dimensional accuracy are further analyzed in the context of the cooling rate effect on the microstructure. The melt pool width control technology improved the cooling rate, whereas the primary dendrite arm spacing (PDAS) and epitaxial growth of large dendrites were suppressed. The continuous evolution of columnar and interlayer equiaxed crystals further improved the thickness uniformity of the thin-walled parts.

References

1. Z.Y. Quan, L. Ge, Z.B. Wei, Y.W. Li, L. Quan, A Survey of powertrain technologies for energy-efficient heavy-duty machinery, IEEE Xplore. 109 (2021) 279-308.
2. J. Brackett, Y.Z. Yan, D. Cauthen, V. Kishore, J. Lindahl, T. Smith, Z. Sudburye, H.B. Ning, V. Kunc, C. Duty, Characterizing material transitions in large-scale Additive Manufacturing, Addi. Manuf. 38 (2021) 101750, https://doi.org/10.1016/j.addma.2020.101750.
3. X.D. Nong, X.L. Zhou, Y.D. Wang. Effects of geometry, location, and direction on microstructure and mechanical properties of 15–5PH stainless steel fabricated by directed energy deposition, Mater. Sci. Eng. A. 821(2021) 141587.
4. I. Jeon, L. Yang, K. Ryu, H. Sohn, Online melt pool depth estimation during directed energy deposition using coaxial infrared camera, laser line scanner, and artificial neural network, Addit. Manuf. 47 (2021) 102295, https://doi.org/10.1016/j.addma.2021.102295.
5. Q. Zhong, X.Y. Tian, X.K. Huang , C.B. Huo , D.C. Li, Using feedback control of thermal history to improve quality consistency of parts fabricated via large-scale powder bed fusion, Addit. Manuf. 42 (2021) 101986, https://doi.org/10.1016/j.addma.2021.101986.
6. Ding D, Pan Z, Cuiuri D, Li H. Wire-feed additive manufacturing of metal components: technologies, developments and future interests. Int J Adv Manuf Technol. 81 (2015) 465–481. https://doi.org/10.1007/s00170-015-7077-3.
7. J. Liu, S. Suslov, Z.C. Ren, Y.L. Dong, C. Ye, Microstructure evolution in Ti64 subjected to laser-assisted ultrasonic nanocrystal surface modification, Int. J. Mach. Tool Manuf. 136 (2019) 19–33, https://doi.org/10.1016/j.ijmachtools.2018.09.005.

8. Z. Li, S. Sui, X. Ma, et al., High deposition rate powder– and wire–based laser directed energy deposition of metallic materials: a review, Int. J. Mach. Tool Manufact. 181 (2022), 103942, https://doi.org/10.1016/j.ijmachtools.2022.103942.
9. P. Kürnstelner, M.B. Wilms, A. Welshelt, et al., High–strength Damascus steel by additive manufacturing, Nature 582 (2020) 515–519, https://doi.org/10.1038/s41586-020-2409-3.
10. R. Mishra, A. K. Sharma, On mechanism of in-situ microwave casting of aluminium alloy 7039 and cast microstructure, Mater. Des. 112 (2016) 97-106, https://doi.org/10.1016/j.matdes.2016.09.041.
11. K. J. Hodder, R. J. Chalaturnyk, Bridging additive manufacturing and sand casting: Utilizing foundry sand, Addit. Manuf. 28 (2019) 649-660, https://doi.org/10.1016/j.addma.2019.06.008.
12. S.Y. Luo, J.N. Yao, J. Li, H. Du, H.Y. Liu, F.P. Yu, Influence of forging velocity on temperature and phases of forged Ti-6Al-4V turbine blade, J. Mater. Res. Technol. 2020; 9(6):12403-12051, https://doi.org/10.1016/j.jmrt.2020.08.106.
13. P. Mandal, H. Lalvani, M. Tuffs, Cold Rotary Forging of Inconel 718, J. Manuf. Process. 46 (2019) 77-99, https://doi.org/10.1016/j.jmapro.2019.08.015.
14. M.J. Hu, K.J. Li, Z.P. Cai, J.L. Pan, A new weld material model used in welding analysis of narrow gap thick-walled welded rotor, J. Manuf. Process. 34 (2018) 614-624, https://doi.org/10.1016/j.jmapro.2018.06.036.
15. H. Cheng, L. Kang, J.C. Pang, B.C. Xue, D. Du, B.H. Chang, Effect of the welding position on weld quality when laser welding Inconel 617 Ni-based superalloy, Opti. Laser. Tech. 139 (2021) 106962, https://doi.org/10.1016/j.optlastec.2021.106962.
16. G.R. Meng, Y.D. Gong, J.D. Zhang, Q.Z. Ren, J.B. Zhao, Microstructure effect on the machinability behavior of additive and conventionally manufactured Inconel 718 alloys, J. Mater. Process. Technol. 324 (2024) 118228, https://doi.org/10.1016/j.jmatprotec.2023.118228.
17. D.F. Peng, H.K. Li, Intelligent monitoring of milling tool wear based on milling force coefficients by prediction of instantaneous milling forces, Mech. Syst. Signal Process. 208 (2024) 111033, https://doi.org/10.1016/j.ymssp.2023.111033.
18. D. Fratila. Evaluation of near-dry machining effects on gear milling process efficiency, J. Clea. Produ. 17 (2009) 839–845, https://doi.org/10.1016/j.jclepro.2008.12.010
19. Y.H. Zhang, W. Cai, Y. He, T. Peng, S. Jia, K. Lai, L. Li, Forward-and-reverse multidirectional turning: A novel material removal approach for improving energy efficiency, processing efficiency and quality. Energy. 260 (2022) 125162, https://doi.org/10.1016/j.energy.2022.125162.
20. N. Asadi, M. Jackson, A. Fundin, Implications of realizing mix flexibility in assembly systems for product modularity—A case study, J. Manuf. Sys. 52 (2019) 13-22, https://doi.org/10.1016/j.jmsy.2019.04.010.
21. X.Z. Chen, C.B. Li, Y. Tang, Q.G. Xiao, An Internet of Things based energy efficiency monitoring and management system for machining workshop, J. Clean. Prod. 199 (2018) 957-968, https://doi.org/10.1016/j.jclepro.2018.07.211.
22. W. Cai , F. Liu , X.N. Zhou , J. Xie, Fine energy consumption allowance of workpieces in the mechanical manufacturing industry, Energy. 114 (2016) 623-633, https://doi.org/10.1016/j.energy.2016.08.028.
23. G.D. Li, M. Reimann, W.H. Zhang, When remanufacturing meets product quality improvement: The impact of production cost, Eur. J. Oper. Res. 271 (2018) 913-925, https://doi.org/10.1016/j.ejor.2018.05.060.
24. C. Tan, F. Weng, S. Sui, Y. Chew, G. Bi, Progress and perspectives in laser additive manufacturing of key aeroengine materials, Int. J. Mach. Tool Manufact. 170 (2021), 103804, https://doi.org/10.1016/j.ijmachtools.2021.103804.
25. B. Bax, R. Rajput, R. Kellet, M. Reisacher, Systematic evaluation of process parameter maps for laser cladding and directed energy deposition, Addit. Manuf. 21 (2018) 487–494, https://doi.org/10.1016/j.addma.2018.04.002.
26. S.K. Everton, M. Hirsch, P. Stravroulakis, R.K. Leach, A.T. Clare, Review of in-situ process monitoring and in-situ metrology for metal additive manufacturing, Mater. Des. 95 (2016) 431–445, https://doi.org/10.1016/j.matdes.2016.01.099.

27. D.K. Kim, W. Woo, J.H. Hwang, K. An, S.H. Choi, Stress partitioning behavior of an AlSi10Mg alloy produced by selective laser melting during tensile deformation using in situ neutron diffraction, J. Alloys Compd. 686 (2016) 281–286, https://doi.org/10.1016/j.jallcom.2016.06.011.
28. ISO/ASTM 52900, Additive manufacturing-General principles-Fundamentals and vocabulary.
29. A. Yadollahi, N. Shamsaei, S. Thompson, D. Seely, Effects of process time interval and heat treatment on the mechanical and microstructural properties of direct laser deposited 316L stainless steel, Mater. Sci. Eng. A. 644 (2015) 171–183, https://doi.org/10.1016/j.msea.2015.07.056.
30. L. Sinclair, S.J. Clark , Y.H. Chen, S. Marussi, S. Shah, O.V. Magdysyuk, R.C. Atwood, G.J. Baxter, M. Jones, D.G. McCartney, C.L. Alex Leung, P.D. Lee, Sinter formation during directed energy deposition of titanium alloy powders, Int. J. Mach. Tool Manufact. 176 (2022) 103887, https://doi.org/10.1016/j.ijmachtools.2022.103887.
31. Z.C. Yang, X.S. Chen, L.D. Zhu, J.S. Ning, Z.Z. Jiang, M. Yu, S.H. Wang, P.S. Xue, P.H. Xu, B. Xin, Comparative study on forming approaches in directed energy deposition of inclined thin-walled structures, Thin-Walled Struct. 191 (2023) 111074, https://doi.org/10.1016/j.tws.2023.111074.
32. R. Weber, et al. Processing constraints resulting from heat accumulation during pulsed and repetitive laser materials processing, Opt. Express 25 (4) (2017) 3966-3979.
33. H.L. Zhang, C. K. P. Vallabh, X.Y. Zhao, Registration and fusion of large-scale melt pool temperature and morphology monitoring data demonstrated for surface topography prediction in LPBF, Addit. Manuf.58 (2022) 103075, https://doi.org/10.1016/j.addma.2022.103075.
34. A. Gaikwad, R. J. Williams, H. Winton, B. D. Bevans, Z. Smoqi, P. Rao, P. A. Hooper, Multi phenomena melt pool sensor data fusion for enhanced process monitoring of laser powder bed fusion additive manufacturing, Mater. Des. 221 (2022) 110919, https://doi.org/10.1016/j.matdes.2022.110919.
35. Q. Guo, M. Qu, L.I. Escano, S.M.H. Hojjatzadeh, Z. Young, K. Fezzaa, L. Chen, Revealing melt flow instabilities in laser powder bed fusion additive manufacturing of aluminum alloy via in-situ high-speed X-ray imaging, Int. J. Mach. Tool Manufact. 175 (2022), 103861, https://doi.org/10.1016/j.ijmachtools.2022.103861.
36. Y.Y. Su, Z.F. Wang, H.F. Lu, K.Y. Luo, J.Z. Lu, Improved wear resistance of directed energy deposited Fe-Ni-Cr alloy via closed-loop controlling laser power, J. Manuf. Process. 75 (2022) 802-813, https://doi.org/10.1016/j.jmapro.2022.01.047.
37. M. H. Farshidianfar, A. Khajepour, A.P. Gerlich, Effect of real-time cooling rate on microstructure in Laser Additive Manufacturing, J. Mater. Process. Technol. 231 (2016) 468-478, https://doi.org/10.1016/j.jmatprotec.2016.01.017.
38. P.A. Hooper, Melt pool temperature and cooling rates in laser powder bed fusion, Addit. Manuf. 22 (2018) 548-559, https://doi.org/10.1016/j.addma.2018.05.032.
39. A. Aggarwal, A. Chouhan, S. Patel, D.K. Yadav, A. Kumar, A.R. Vinod, K.G. Prashanth, N.P. Gurao, Role of impinging powder particles on melt pool hydrodynamics, thermal behaviour and microstructure in laser-assisted DED process: A particle-scale DEM – CFD – CA approach, Int. J. Heat Mass Transf.158 (2020) 119989, https://doi.org/10.1016/j.ijheatmasstransfer.2020.119989.
40. Z. Sun, W. Guo, L. Li, In-process measurement of melt pool cross-sectional geometry and grain orientation in a laser directed energy deposition additive manufacturing process, Opt. Laser Technol. 129 (2020) 106280, https://doi.org/10.1016/j.optlastec.2020.106280.
41. Q. Wang, P. Michaleris, A. Nassar, J. Irwin, Y. Rena,C. Stutzman, Model-based feedforward control of laser powder bed fusion additive manufacturing, Addit. Manuf. 31 (2020) 100985, https://doi.org/10.1016/j.addma.2019.100985.
42. B. Zheng, Y. Zhou, J.E. Smugeresky, J.M. Schoenung, E.J. Lavernia, Part II experimental investigation and discussion, Metall. Mater. Trans. A 39A (2008) 2237–2245.
43. Y.Y. Su, Z.F. Wang, X. Xu, K.Y. Luo, J.Z. Lu, Effect of closed-loop controlled melt pool width on microstructure and tensile property for Fe-Ni-Cr alloy in directed energy deposition, J. Manuf. Process. 82 (2022) 708–721, https://doi.org/10.1016/j.jmapro.2022.08.049.

44. W.W. Deng, C.Y. Wang, H.F. Lu, X.K. Meng, Z. Wang , J.M. Lv ,K.Y. Luo, J.Z. Lu, Progressive developments, challenges and future trends in laser shock peening of metallic materials and alloys: A comprehensive review, Int. J. Mach. Tool Manufact. 191 (2023) 104061.
45. H. Zhang, D.H. Dai, L.H. Yuan, H. Liu, D.D. Gu, Temperature gradient induced tough-brittle transition behavior of a high-strength Al-4.2Mg-0.4Sc-0.2Zr alloy fabricated by laser powder bed fusion additive manufacturing, Addit. Manuf. 73 (2023) 103655.
46. G.Q. Dai, Z.G. Sun, Y.S. Li, J. Jain, A. Bhowmik, J.J. Shinjo, J.Z. Lu, C. Panwisawas, Grain refinement and columnar-to-equiaxed transition of Ti6Al4V during additive manufacturing via different laser oscillations, Int. J. Mach. Tool Manufact.189 (2023) 104031.
47. B. Zheng, Y. Zhou, J. Smugeresky, J. Schoenung, E. Lavernia, Thermal behavior and microstructure evolution during laser deposition with laser-engineered net shaping: part II. experimental investigation and discussion, Metall. Mater. Trans. A. 39 (2008) 2237 -2245.
48. M. Sudmanns, J. A. El-Awady, Connecting processing conditions to microstructure evolution in laser powder bed fusion via defect modeling, Scr. Mater. 245 (2024) 116035.
49. A. Abdali, S. H. Nedjad, H. H. Zargari, A. Saboori, M. Yildiz, Predictive tools for the cooling rate-dependent microstructure evolution of AISI 316L stainless steel in additive manufacturing, J. Mater. Process. Technol. 29 (2024) 5530-5538.
50. H.S. Carslaw, J.C. Jaeger. Conduction of Heat in Solids 1947, second ed., Oxford University Press, Oxford, 1959 (Oxford, at the Clarendon Press).
51. N.N. Rykalin, Calculations of thermal processes in welding, Mashgiz, Moscow, 1960.
52. Z.D. Shang, Z.R. Liao, J.A. Sarasua, J. Billinghamd, D. Axinte, On modelling of laser assisted machining: Forward and inverse problems for heat placement control, Int. J. Mach. Tool Manufact. 138 (2019) 36-50, https://doi.org/10.1016/j.ijmachtools.2018.12.001.
53. C. Hagenlocher, P. O'Toole, W. Xu, M. Brandt, M. Easton, A. Molotnikov, Analytical modelling of heat accumulation in laser based additive manufacturing processes of metals, Addit. Manuf. 60 (2022) 103263.
54. R. Weber, T. Graf, P. Berger, V. Onuseit, M. Wiedenmann, C. Freitag, A. Feuer, Heat accumulation during pulsed laser materials processing, Opt. Express 22 (9) (2014) 11312–11324.
55. Y.Y. Su, G. Xu, X. Xu, K.Y. Luo, J.Z. Lu, In-situ thermal control-assisted laser directed energy deposition of curved-surface thin-walled parts, Addit. Manuf. 83 (2024) 104061.

Chapter 6
Effect of the Melt Pool In-Situ Control on the Mechanical Properties of Laser Directed Deposition 316L Stainless Steel Thin-Walled Parts

6.1 Introduction

With the upgrading of modern manufacturing, heavy equipment is developing rapidly in the direction of large-scale, high-performance and integration, which puts forward higher requirements for manufacturing technology. Large components such as hardware stamping dies, space shuttle turbine blades and hydraulic support columns for coal machines are subject to environmental corrosion damage on their surfaces, leading to scrap failure of the entire component [1–3]. According to ISO/ASTM 52900:2021, additive manufacturing is the process of joining materials to make parts from 3D model data, usually layer upon layer, as opposed to subtractive manufacturing and formative manufacturing methodologies. And directed energy deposition, as an additive manufacturing process in which focused thermal energy, is used to fuse materials by melting as they are being deposited. In this study, the manufacturing process can be defined as DED-LB/M/AISI 316L. The process category is directed energy deposition (DED), a laser beam is used for consolidation (-LB), and it is a metallic material (/M), more specifically AISI 316L (/AISI 316L). Due to the advantages of "reduced machining procedures, shorter cycle times and rapid manufacturing", DED technology has been widely used in petroleum, aerospace, automotive and marine applications [4, 5].

Complex aerospace parts manufactured using LDED, particularly thin-walled parts, are typically scanned along a single multilayer path. Continuous deposition, i.e., no cooling time set between layers, is typically used for productivity reasons. This leads to the problem of thermal accumulation effects because the component is not cooled in time. Excessive accumulation of heat influences the forming quality of components, including the dimensional accuracy [6, 7], surface roughness [8, 9], interlayer bonding [10, 11], and overall performance [12–14]. Among them, metallurgical defects seriously affect the corrosion resistance. Defect formation is related to the complex and variable heat transfer patterns and dynamic characteristics of the

K. Luo et al., *Melt Pool In-Situ Controlling of Laser Additive Manufacturing Technology*, https://doi.org/10.1007/978-981-95-4597-1_6

melt pool during processing. As an example, lack of fusion (LOF) pores are common defective features in additive manufacturing which strongly affect corrosion resistance [15–17]. Schaller et al. [18] showed that when deposited 304L, LOF pores generally act as preferential damage points in the corrosive solution due to their irregular geometry. Melia et al. [19] investigated the effect of LOF pores and δ-ferrites on the corrosion performance in different heat input deposition fabrication of stainless steel. In addition, Lee et al. [20] investigated the changes in the physical phase of titanium alloys by setting up different interlayer cooling times, which provided guidance for the improvement of the mechanical properties. However, this approach increases the interlayer processing time and reduces productivity. Qiu et al. [21] prepared a Ni-based Hastelloy C276 alloy using cold metal transfer-based LDED to reduce the mechanical anisotropy of the alloy via interlayer temperature control. This method is commonly used for the preparation of dissimilar alloys and for improving material properties via laser-arc composite fabrication; however, it is not suitable for solving the problem of thermal effects in conventional materials.

Defect formation is related to the complex and variable heat transfer patterns and dynamic characteristics of the melt pool during processing. To reduce the effect of thermal effects during deposition, changes in laser process parameters are usually used [12]. Xu et al. [22] compared the microstructure and corrosion resistance of Cr-rich SS coatings fabricated by conventional and extreme high- speed laser metal deposition, and demonstrated that increasing the scanning speed helped to reduce porosity and improve corrosion resistance. Yang et al. [14] investigated the effect of interlayer cooling time on the thermal accumulation effect and established the link between interlayer cooling time and forming quality. Farshidianfar et al. [23, 24] proposed a method to vary the scanning speed to ensure the cooling rate and investigated its effect on the organisation of 316L stainless steel prepared by additive manufacturing.

To cope with the thermal effect problem, changing laser process parameters often requires extensive exploratory orthogonal experiments and optical experimental validation to finalize the appropriate parameters. This off-line inspection-verification approach not only increases time and material costs, but also does not address the fundamental problem from the heat source perspective. The high-energy laser beam melts the substrate and powder to form a melt pool. Real-time monitoring and online control of the melt pool during deposition is crucial, it is also important to solve the thermal effect problem [25–27]. Among them, the melt pool size significantly affects the forming accuracy, and the monitoring and control of the melt pool geometry has become a popular research topic [12]. Vasinonta et al. [28] established a process mapping of melt pool sizes to investigate effects of melt pool aspect ratio on deposition quality. Heralic et al. [29] used camera monitoring and iterative learning method to control the width and height of the deposited layers by varying the wire feeding speed of each layer. Ding et al. [25] used an infrared imaging device to collect melt pool images, and a photoelectric sensor and a PID controller to detect and control the powder flow rate to improve the uniformity of melt pool dimensions.

This chapter focused on the closed-loop control of surface finish, pore defects and corrosion properties of 316L stainless steel deposited by DED.A coaxial infrared camera was used to acquire images and extract melt pool widths. The melt pool width was adjusted online using PID parameters to converge to a set value and ensure its uniformity. The roughness, pore distribution and corrosion performance of deposited specimens in the traditional and control modes were compared. The positive impact of the closed-loop controlled melt pool width process on improving surface finish, reducing pore defects and improving corrosion performance was analyzed.

6.2 Experiment and Characterizations

6.2.1 LDED and Melt Pool In-Situ Control Process

Figure 6.1 presented the DED system with closed-loop control system. As shown in Fig. 6.1a, it consisted of a fibre-coupled diode laser system, KUKA robot, CNC equipment, powder feeder, 3D printing head, gas bottle, infrared camera and a PC with built-in melt pool width control software. During the deposition process, 45 steel and AISI 316L austenitic stainless steel were used as the substrate material and molten powder material, respectively. The chemical compositions were shown in Table 6.1. As shown in Fig. 6.1b, several dimensions of 68 mm × 25 mm × 4 mm (length ×height× thickness) thin-walled parts were deposited layer by layer on the 45 steel substrate in a "S" scanning path. Taking into account the forming of the specimen and reading the literature, the initial power was set to 2000 W. Except for the laser power, the deposition parameters (as shown in Table 6.2) were kept constant to set the control group.

As the key role in achieving closed-loop control of the width, the selection and construction of the infrared camera (Fig. 6.1c) was critical. The infrared camera was placed coaxially on the 3D printing head to facilitate the acquisition of a live image of the melt pool from an overhead perspective, and an infrared filter was installed on the front of the camera chip. The camera's external interfaces included an I/O port, a water cooling port, a head connector, and an Ethernet port. The water cooling interface was connected to a water cooling circuit to protect the unit from overheating. The camera was co-axially installed on the laser head via the connector, allowing for manual focus. There were several steps to get the melt pool width: image greyscale processing, average filtering, denoising, threshold segmentation and smoothing of the profile [26, 30]. The acquisition frequency and resolution of the camera were 50 fps and 720 × 540 pixels, respectively. The frame rate used in the monitoring is 302.3 fps. Basic image pre-processing methods were used to obtain the melt pool image as shown in Fig. 6.1b and the melt pool width was obtained by simple image pixel calculation.

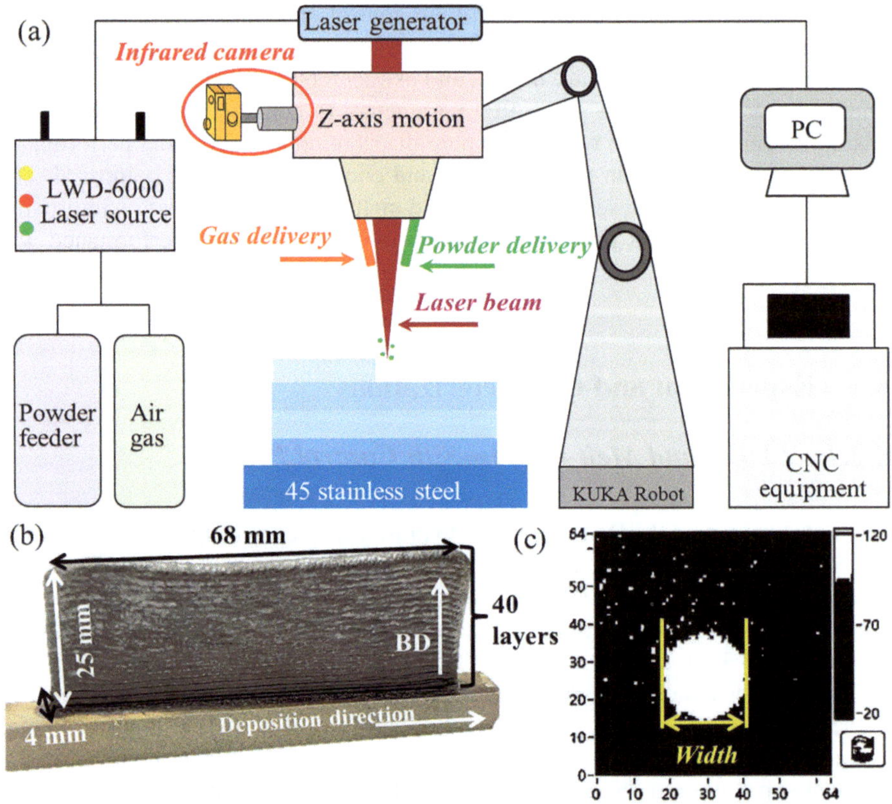

Fig. 6.1 The directed energy deposition equipment with closed-loop control feedback system. (**a**) Visual representation, (**b**) thin-walled part fabricated by DED, (**c**) melt pool width extracted by image processing

Table 6.1 The chemical compositions of 45 steel and 316L stainless steel

Element (wt. %)	Ni	Cr	Mn	Si	C	Mo	Fe
45 steel	/	/	1.34	2.77	0.9	/	Bal.
AISI 316L austenitic stainless steel	12–13	16–18	0.2	0.8	0.03	2–3	Bal.

Table 6.2 The laser processing parameters in DED

Parameters	Value	Units
Laser power	2000	W
Powder feed rate	72	g/min
Scanning speed	10	mm/s
Overlapping rate	50%	/
Spot diameter	4	mm

Figure 6.2 illustrated the real-time melt pool image and single track morphology in both modes. An infrared camera mounted above the laser head acquired the image of the melt pool during deposition. In the traditional mode, due to certain heat accumulation (high heat input and slow cooling rate), the irregular melt pool shape was collected (Process I). The increased melt pool width further affected the uniformity and flatness of the deposited layer (Process II). PID method was introduced in the control mode (Process III). The difference between the real-time melt pool width and the set value was set as an input to the control process, and the laser power was adjusted to improve the uniformity of the melt pool width. The average melt pool width of the first 3–5 deposited layers was generally set to a set value. The parameters K_p, T_i and T_d were determined by the engineering neaten method according to the oscillatory behavior of the melt pool width curve. The effective control of the melt pool width was fed back to the camera field of view with a regular melt pool shape and a stable width. The deposited layers in the controlled mode were flat and uniform (process IV), and the DED system prepared thin-walled parts by layer-by-layer stacking, thus forming a closed loop process (process V).

The PID control played a key role in the feedback control of the melt pool width. PID control gains were arrived at by empirical tuning on the live system. The actual parameter tuning should be changed at any time according to the actual situation. When the controlled physical quantity oscillates near the target value, K_i should be increased; if the oscillation still exists, K_P should be reduced appropriately; if the controlled quantity deviates from the target value greatly, K_p should be increased, K_i should be reduced or K_d should be increased. Most of the noise was filtered using a low-pass filter to eliminate the interference on the derivative control. The integration time was set to 800 μs, and the sensor gain was adjusted to increase the signal strength. The aforementioned approaches resulted in a more stable control system output. In the preliminary stage of the closed-loop mode, K_p accelerated the system response, while significantly suppressing the melt-pool width. Consequently, K_i and K_d were enhanced, thereby effectively reducing the width suppression and slowing down the oscillations. In addition, this method can improve the control stability before adjusting the melt pool to reach the target width. During deposition, the three-component parameters were adjusted based on the fluctuation range of the standard melt-pool width, which was determined as $K_p = 50$, $T_i = 800$, and $T_d = 0.05$.

6.2.2 Forming Quality Characterization and Performance Testing

The surface morphologies and roughnesses of the thin-walled parts were measured using a Leica TCS SP5 laser confocal microscope. The internal pore distribution and size statistics of the specimens were measured by an AX-3000CT device with a pore detection accuracy of 2 μm. The samples were tested on a three-electrode electrochemical workstation (CHI660e) to investigate the corrosion behaviour of the

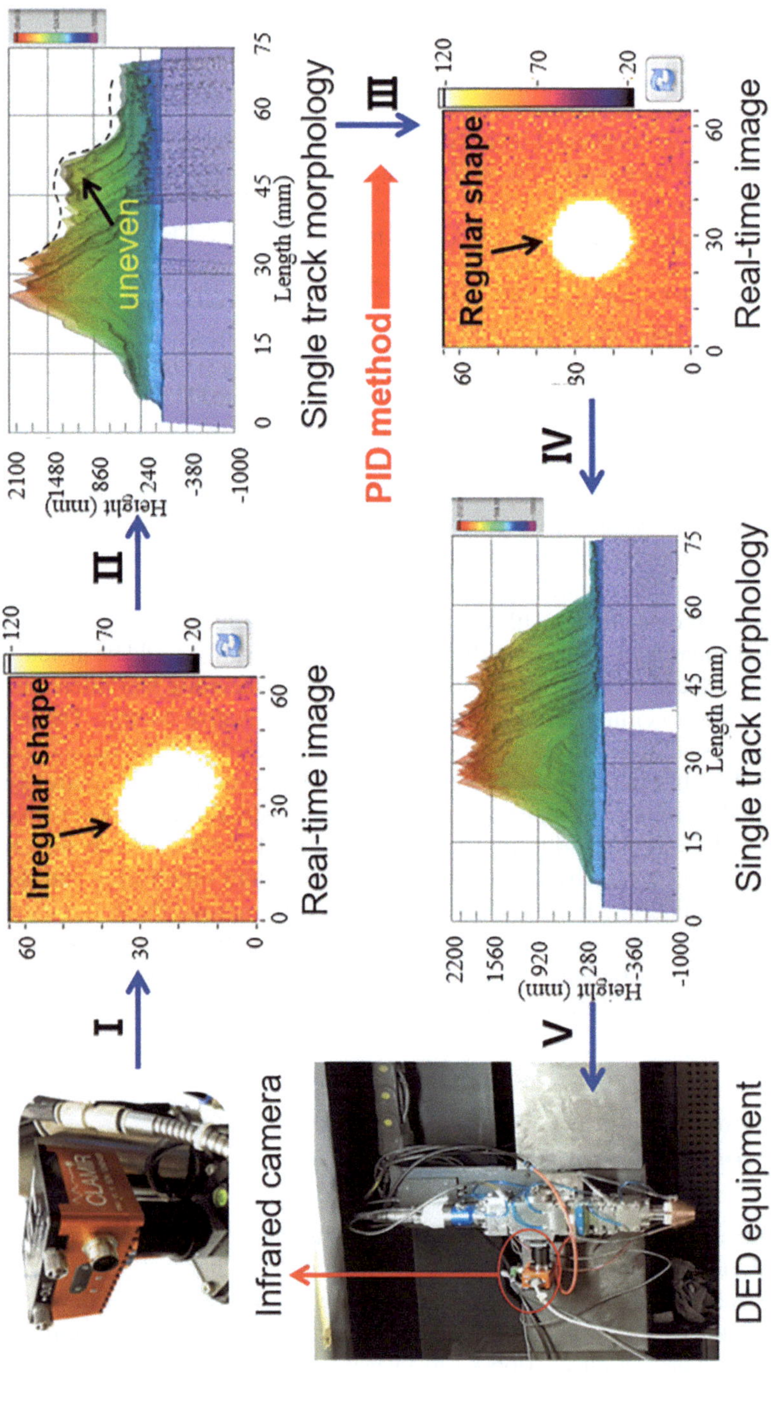

Fig. 6.2 Closed-loop control of melt pool width using PID technology in DED

Fe-Ni-Cr coating. The experiments were carried out using the 3.5 wt% NaCl solution with a contact area of approximately 25 mm^2 at room temperature. In addition, the saturated calomel electrode and the aluminium foil sheet were used as reference and auxiliary electrodes, respectively. The open circuit potential (OCP) was obtained prior to testing. Each sample was immersed for 20 min before characterising the electrochemical impedance spectra (EIS) under OCP. A stable frequency range of 10–2 to 105 Hz with voltage amplitude of 5 mV was used in the experiment. Three control groups were set up for each test. The tensile experiments were performed on a Zwick Z2.5 TH tensile testing machine equipped with a laserXtens tensiometer. The tensile experiments were performed on three specimens per group.

6.3 Control Results and Forming Quality Characterization

6.3.1 Melt Pool Control Results

Figure 6.3 shows the laser power and melt pool width curves for depositing thin-walled parts in both modes, with a total of 40 layers. In the traditional mode, it is observed that the melt pool width gradually increases from 2.3 mm to ~2.75 mm. In the control mode, the first five layers of the thin-walled part are deposited with constant power, and the average width is set to the set value. Closed-loop control is introduced at the deposition of the sixth layer, and the difference between the set

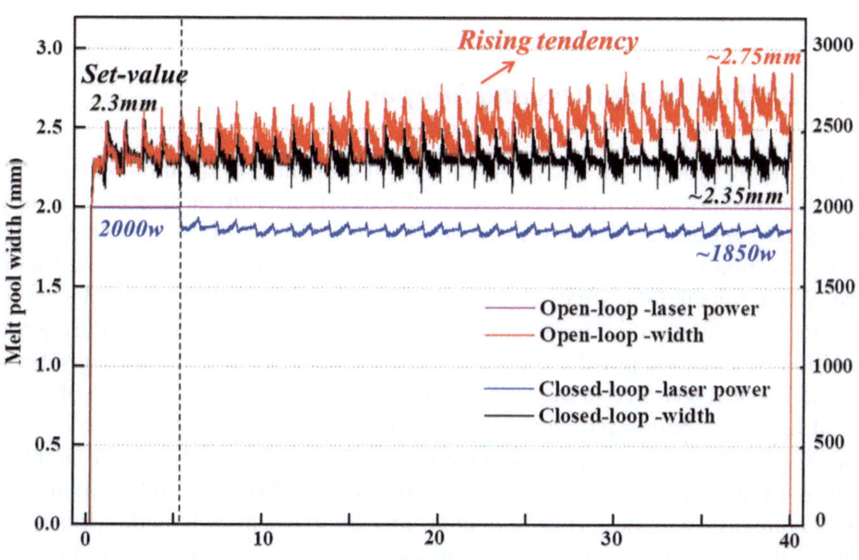

Fig. 6.3 Laser power and melt pool width curves of thin-walled specimen in the traditional and control mode

value and the real-time width is used as the input of PID control. The controller adjusts the parameters according to the curve oscillation changes, and sends a signal to adjust the laser power, so that the laser power curve shows dynamic changes. Eventually, the laser power reduces from 2000 W to ~1850 W, with stable melt pool width at ~2.35 mm.

It can be noticed that the width curves in both modes have obvious rising peaks that appear regularly, which are determined by the deposition strategy. The deposition of thin walls is performed using alternating scans, where the end point of each layer is also the starting point of the next layer, so the melt pool width increases abruptly at both ends, producing sharp peaks. In addition, the presence of a downward spike at the end of each layer deposition was also found in the control mode. Because there is a certain thermal accumulation at the end of the first five layers of deposition and the melt pool width already has a tendency to gradually increase. After the introduction of PID control, the controller detects the signal of increasing melt pool width. The command to reduce the input power is then sent to the laser. A gradual decrease in the melt pool width profile can be observed after the power reduction. To maintain thermal equilibrium, the laser power is gradually increased at the end of a single deposition. Thus, the laser power and melt pool width curves are characterized by "decrease - gradual increase - decrease". The time nodes of power and width change do not occur at the same moment, and the melt width always lags behind the laser power. This is related to the time consumed to collect and control the melt pool width [26, 28]. Therefore, in the actual addition process, the time node of the laser power emitted by the laser is smaller than the time node when the width value ends after the melt pool generation.

6.3.2 *Surface Roughness and Morphology*

Figure 6.4 shows the 3D surface morphology and roughness of the thin-walled parts in both modes. As shown in Fig. 6.4a, b, the differences in surface morphology between two modes are mainly in the middle and top of the thin-walled parts, because the first five layers are deposited with constant power in the closed-loop mode. When selected the areas where the differences exist in the two modes separately, it can be found that due to some of the powder not being melted during the deposition process, the powder is adhered to the specimen surface, forming spherical particles of larger size in the traditional mode (Fig. 6.4a_1). There is no obvious trace of interlayer build-up on the specimen surface which is not flat. In the control mode (Fig. 6.4b_1), the specimen surface has a distinct and well-defined lamellar morphology, and the number of spherical particles formed is small and small in size. Further comparison of the corresponding 3D morphologies in both modes (Fig. 6.4c, d) shows that the flatness and finish of the specimen surface in the control mode is significantly higher. Figure 6.4e shows the surface roughness results in both modes, and it can be seen that the roughness values are close at the bottom of the thin-walled parts. However, in the middle and top of the deposition, the roughness in the

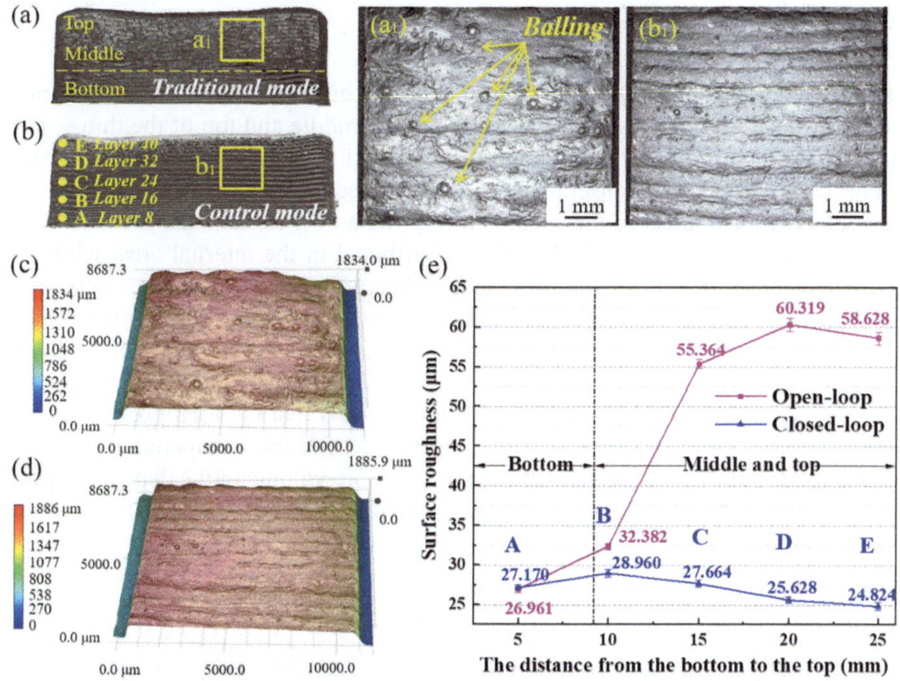

Fig. 6.4 Surface finish and roughness of thin-walled parts in the traditional mode (**a**, **a**$_1$, **c**) and control mode (**b**, **b**$_1$, **d**). (**a** and **b**) Thin-walled specimens deposited in both modes, (**a**$_1$ and **b**$_1$) surface morphology of the selected middle and top areas of specimens, (**c** and **d**) 3D morphological characteristics of the selected area, (**e**) surface roughness statistics in both modes

traditional mode shows a continuous increasing trend with a maximum roughness of 60.319 μm. On the contrary, the roughness of the specimen surface in the control mode remains in a stable variation range with an average roughness of 26.850 ± 1.250 μm. The closed-loop control of the melt pool width results in a maximum reduction of 57.5% in surface roughness.

The surface roughness of thin-walled parts is related to the temperature coefficient of melt pool surface tension [31]. The main form of convection in the melt pool is the double-ring vortex flow dominated by the Marongoni flow [32]. In the open-loop mode, the negative Marongoni flow pushes upward to the central part of the melt pool due to increased thermal accumulation, causing the low curvature angle of the melt pool to increase with increasing linear energy (heat vs. scanning speed). The liquid metal is under tension and the velocity is increasing, in order to reach equilibrium at the position of maximum tension, a certain scale of sludge is bound to form, thus forming a convex morphology. In addition, the unmelted powder particles gather at the wall surface, producing a more or less necking effect, which leads to surface roughening. In the closed-loop mode, the reduced melt pool thermal gradient reduces the centrifugal Marangoni flow, which further limits the lateral expansion of the melt pool, resulting in a more polished surface.

6.3.3 Porosity Analysis

Figure 6.5 depicts the 3D reconstruction maps of internal defect features and pore statistics results of the regions selected from the middle and top of the thin-walled part in two modes. It can be seen from Fig. 6.5a, b that the pore distribution is sparse in the control mode and denser in the traditional mode. Comparison of the pore sizes for selected local areas of the specimen (Fig. 6.5a_1, b_1) reveals that the traditional mode mostly has obvious block defects distributed in the internal area, while the pore defects are much smaller after control. According to the pore size statistics of specimens, it can be found that the number of pores (384) is 6.6% lower than that in the traditional mode (411). The pore size in the traditional mode is mainly in the range of 10–20 μm, larger than in the control mode (5–10 μm). Further comparison of the pore volumes in both modes (Fig. 6.6) reveals that the total pore volume in the control mode is 106.16 × 10^4 μm^3, which is 52.5% less compared to the traditional mode (μm^3). Among them, the maximum pore volume of the deposited specimens in the traditional mode reaches 18 × 10^4 μm^3, while it is 11 × 10^4 μm^3 in the control mode.

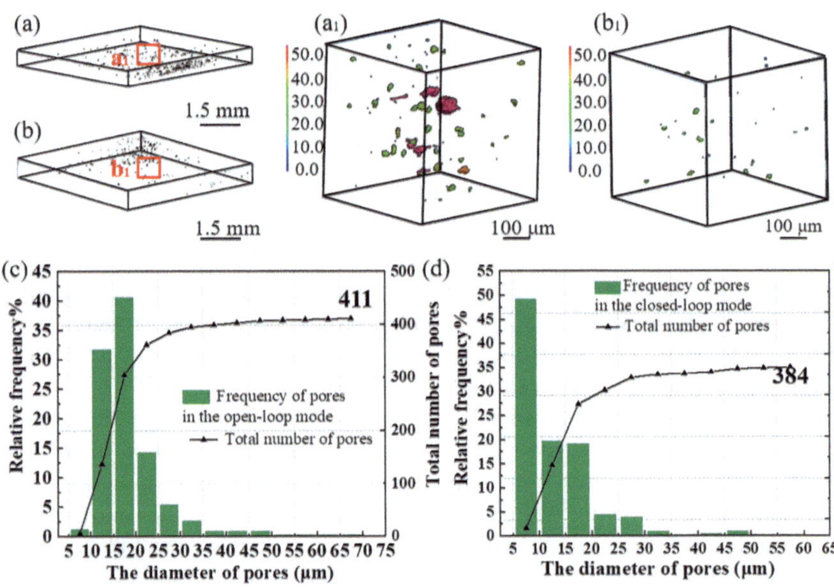

Fig. 6.5 3D pore defects and number distribution statistics of thin-walled parts in the traditional mode (**a**, **a_1**, **c**) and control mode (**b**, **b_1**, **d**). (**a** and **b**) Overall pores in thin-walled parts sheet, (**a1** and **b1**) pore distribution in selected local cubes, (**c** and **d**) number statistics of pores

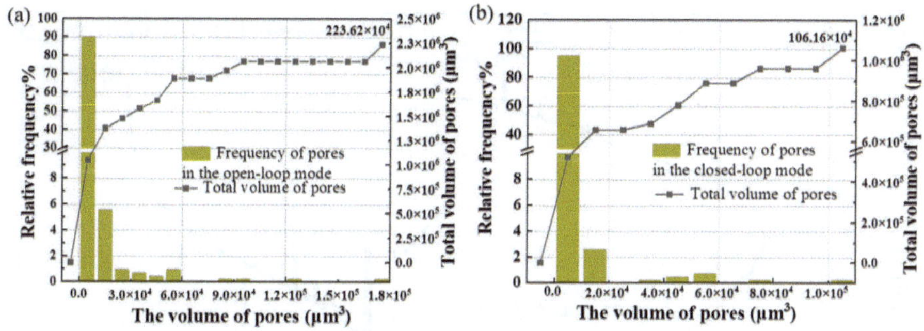

Fig. 6.6 Volume statistics of pores in the traditional (**a**) and control (**b**) modes

6.4 Corrosion Resistance

6.4.1 Potentiodynamic Polarization Test

Figure 6.7 depicts the typical polarization curves of the thin-walled part specimens tested in 3.5 wt% NaCl solution in both modes, where the presence of distinct passive regions can be clearly seen. As shown in Fig. 6.7a, compared to the traditional mode, the polarization curve in the control mode shifts noble direction. The current density of the sample decreases significantly, indicating that the closed-loop control has a positive effect on improving corrosion resistance. Figure 6.7a_1 reveals that there is a current density fluctuation in the traditional mode, which indicates the metastable pit growth [33]. No significant current density fluctuations are found in the control mode, indicating that the specimen surface did not produce metastable pits. Figure 6.7a_2, a_3 also shows the corrosion pits on the surface of the specimen after the polarization test in both modes. The number of corrosion pits in the control mode is significantly reduced than the traditional mode, with sparser distribution.

Table 6.3 shows the electrochemical parameters obtained from the polarization curves. The data indicates that the closed-loop control process has a significant benefit in improving the corrosion resistance of 316L stainless steel. Compared to the traditional mode, a more stable passive plateau exists in the control mode. The performance of the passive film also seems to be improved by closed-loop control, leading to a decrease of I_{pass} (at the potential of 0 V) from 20.58 μA/cm^2 (S-Traditional) to 12.85 nA/cm^2 (S-Control). In addition, the control sample also exhibits a large anodic shift of the Epit, from 0.430 V (traditioanl mode) to 0.512 V, further demonstrating that controlling the melt pool width and heat input can significantly suppress pit initiation and improve corrosion resistance.

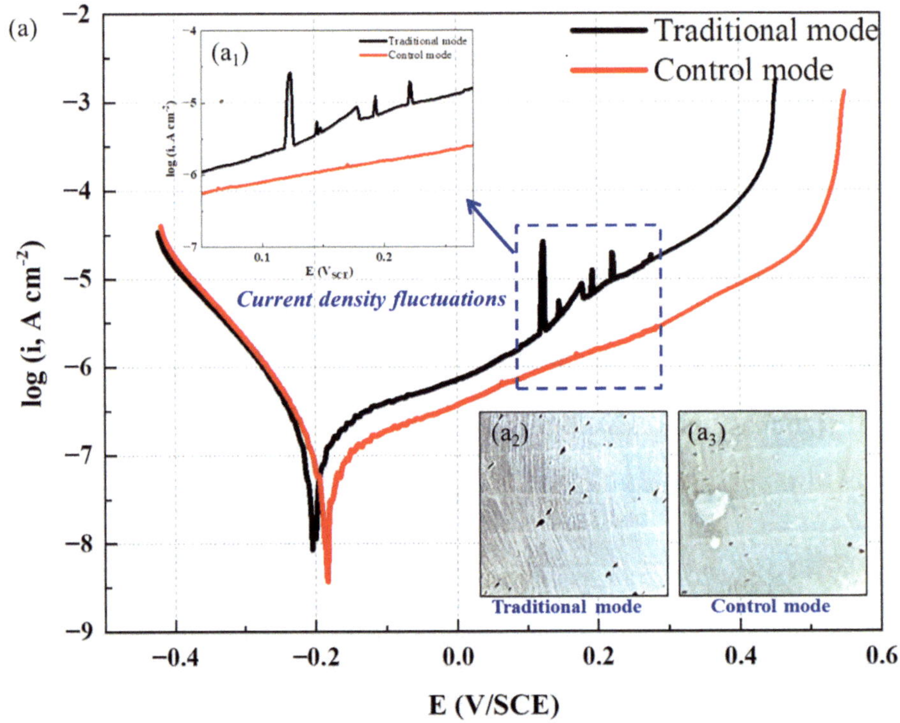

Fig. 6.7 Potentiodynamic polarization diagrams of 316L stainless steel fabricated in different modes in 3.5 wt% NaCl solution (a) Potentiodynamic polarization diagrams, (a_1) amplification of local dynamic curves, (a_2) corrosion pit morphology in the traditional mode, (a_3) corrosion pit morphology in the control mode

Table 6.3 The electrochemical parameters for potentiodynamic polarization curves of the specimens in two modes in 3.5 wt% NaCl solution

Sample	$I_{pass(at\ 0\ V)}$ (μA/cm^2)	E_{pit} (V_{SCE})
S-Traditional	20.58 ± 4.5	0.430 ± 0.005
S-Control	12.85 ± 5.0	0.512 ± 0.010

6.4.2 Corrosion Morphology

Figure 6.8 shows the shape of surface corrosion pits and geometric measurement results after the corrosion behavior test in both modes. In the traditional mode, the shape of the pits produced on the specimen surface are mostly elongated (Fig. 6.8a), while the control mode is mainly oval-like pits (Fig. 6.8c). When characterize the pit depths and widths separately along the direction of the arrow shown in figures, it can be found that the depth of a single pit in the traditional mode (Fig. 6.8b) is uniform, with two crater depths of −10.80 μm and −22.68 μm, respectively. In the control mode, the dynamic changes of the pits curve indicates the uneven depth of the corrosion pits, with the deepest depth of −20.42 μm, which is also smaller than in the traditional mode. The shallowest pit is only −5.83 μm. The large number of corrosion pits on the coating surface indicates that the main types of corrosion include localized and pitting corrosion [34].

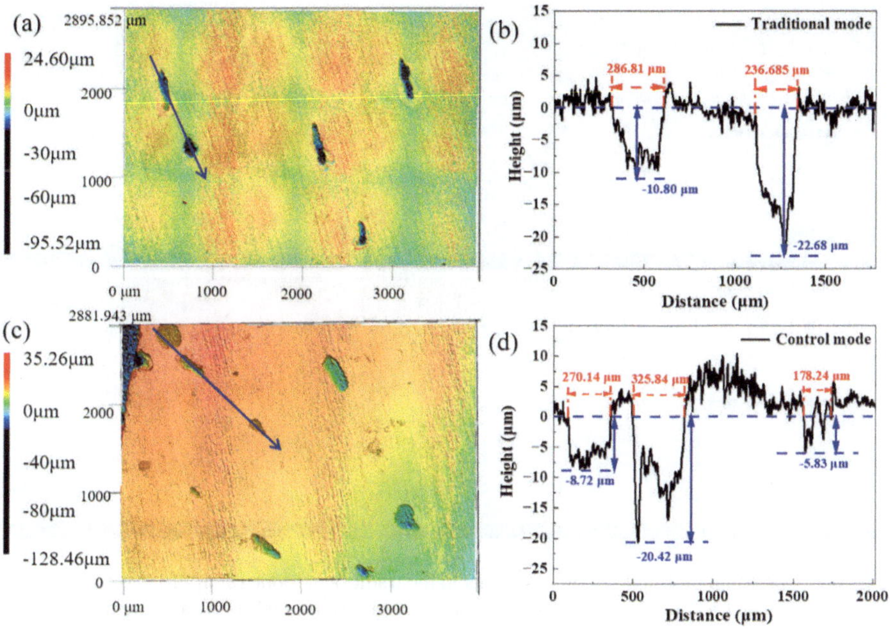

Fig. 6.8 Corrosive surface morphologies and measuring results of corrosion pits. (**a** and **b**) Traditional mode, and (**c** and **d**) control mode

In the control mode, the size of corrosion pits is generally smaller than in the traditional mode, and the shape is mostly circular or elliptical (Fig. 6.9b_1, b_2). Analysis of individual corrosion pits shows that, in addition to pitting pits, many nanoscale Fe-rich phases and corrosion exfoliation of larger sizes are found around them. As in Fig. 6.9a, the corrosion behavior in the control mode is also that intergranular corrosion occurs after further expansion of pitting corrosion. In addition to the regularly arranged dendrites seen inside the grains, exfoliation is also found inside the corrosion pits. With the closed-loop control process, the diffusion behavior of the C element in the material to the grain boundaries is suppressed by the reduced heat input, thus reducing the production of C-Cr compounds. According to the Cr-depleted theory, intergranular corrosion is suppressed by the decrease of Cr element content at grain boundaries and the increase of intergranular Cr elements.

6.4.3 Electrochemical Impedance Spectra

The Nyquist plots, the Bode-impedance, and Bode-phase plots of the specimens in two modes are shown in Fig. 6.10. As shown in Fig. 6.10a, the capacitance ring radius of the coating in the control mode is clearly larger than that of the coating in the traditional mode, indicating that the passive layer formed on the surface of the coating in the control mode presents superior protection. The Nyquist plot is

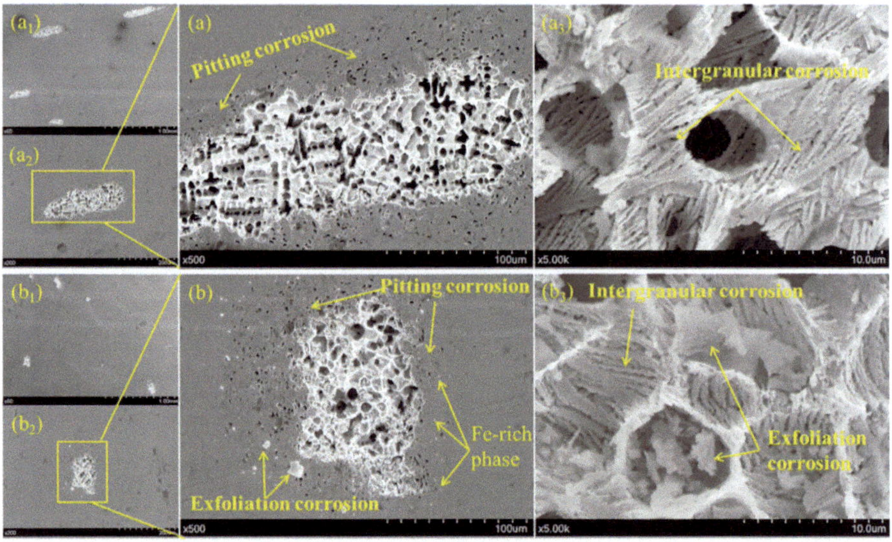

Fig. 6.9 Corrosion morphology of specimens in the traditional mode (**a**, **a**$_1$–**a**$_3$) and control mode (**b**, **b**$_1$–**b**$_3$). (**a** and **b**) individual corrosion pit profiles, (**a**$_1$, **a**$_2$, **b**$_1$, and **b**$_2$) small magnification corrosion profiles, (**a**$_3$ and **b**$_3$) magnified intergranular corrosion features

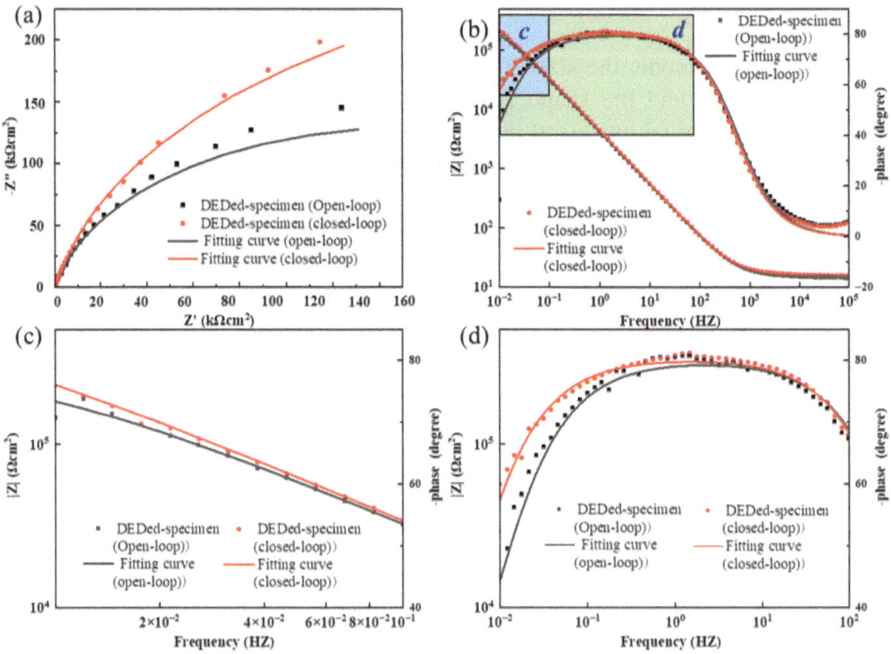

Fig. 6.10 EIS measurements of specimens in the traditional and control modes. (**a**) Nyquist plots, (**b**) Bode modulus plots and Bode-phase plots (**c**), Bode modulus plots at low frequency, (**d**) Bode plots in the mid-frequency range

Table 6.4 The fitted electrochemical parameters for impedance spectra of specimens in the traditional and control mode

Sample	$Q_{dl}(\mu\Omega^{-1}cm^{-2}s^{n})$	$R_s/\Omega{*}cm^2$	$R_{ct}/\Omega\ cm^2$	n_{dl}	$X^2(\times10^{-3})$
S-Traditional	45.36 ± 3.0	1.50 ± 0.20	(3.090 ± 0.10) × 10^5	0.8893 ± 0.03	5.319 ± 0.50
S-Control	43.69 ± 3.5	1.64 ± 0.30	(5.872 ± 0.10) × 10^5	0.8937 ± 0.01	1.278 ± 0.40

characterized by incomplete semicircles, which indicates that the charge transfer process plays a major role in the corrosion process [35, 36]. Figure 6.10b presents Bode-impedance and Bode-phase plots, the approximation between log|Z| and log (frequency) confirms to a linear function. When focus at low frequency, it can be found that the impedance values of the specimens in the control mode are larger than in the traditional mode (Fig. 6.10c). The Bode plots of the coatings in both modes exhibit a broad peak in the mid-frequency range with a phase of 70°–80° (Fig. 6.10d) [37, 38]. Larger phase angle values are detected over a wider frequency range for the specimen in the control mode, indicating a more stable passive film [39].

As shown in Table 6.4, the charge-transfer resistance (R_{ct}) is generally proportional to the corrosion resistance of the passive film [40]. Compared to 3.090 × 10^5 Ω cm^2 in the traditional mode, the specimen achieves an R_{ct} value of 5.872 × 10^5 Ω cm^2 in the control mode, indicating better corrosion resistance. In addition, the results for the χ^2 in all states range between 1.278 and 5.319 × 10^{-3}.

Figure 6.11 illustrates the results of XPS analysis in the passivated layer in two modes. Both specimen surfaces are readily oxidized to form Cr and Fe passivation products. Figure 6.11a, c confirm the presence of Cr_2O_3 in the surface films in both modes. The existence of FeO and Fe_2O_3/FeOOH is demonstrated after the Fe $2p_{3/2}$ spectra (Fig. 6.11b, d) [41]. The passive films formed on the specimens prepared in both modes are mainly Cr^{3+}-rich oxides/hydroxides. However, the Cr cation fraction on the surface of the specimens in the control mode (~85.34%) is significantly higher than that in the traditional mode (~70.44%), indicating an increase in the oxidation kinetics of Cr on the surface of the specimens after control. According to the electrochemical test results, due to the enrichment of Cr_2O_3 and Fe_2O_3/FeOOH promoting the formation of a dense oxide film and improving the stability of the passive film, the better corrosion resistance of the specimen surface is improved in the control mode [42, 43].

6.4.4 Corrosion Mechanism Analysis

Pore density plays a key role in affecting the corrosion resistance of the deposited specimens [34, 44]. In the traditional mode, the current density fluctuations in the polarization curve (Fig. 6.7) may be caused by the presence of large pores on the specimen surface, which affects the integrity of the formed passivation film. Due to the existence of pores, the passivation layer is weak, and the metal is in contact with

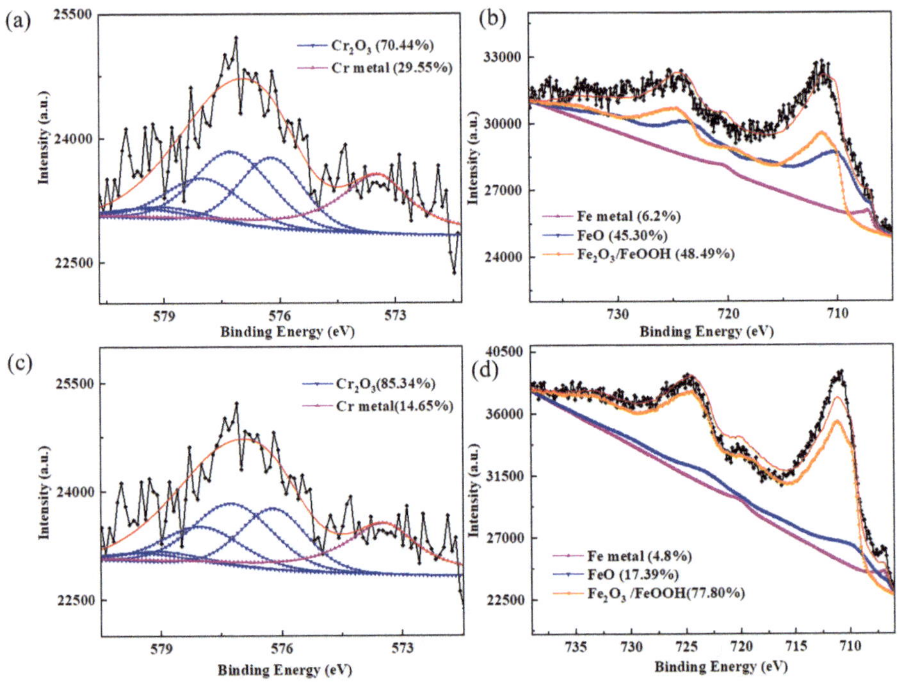

Fig. 6.11 XPS analysis of $Cr2p_{3/2}$ and $Fe2p_{3/2}$ in the passive film formed on surfaces of samples in 3.5 wt% NaCl solution. (**a** and **b**) Traditional mode, and (**c** and **d**) control mode

the corrosion medium to form an active center, the size of the active center increases as the corrosion proceeds, resulting in pitting corrosion. The presence of pores will increase the contact area between the corrosive medium and the specimen surface, further increasing the corrosion extension rate [45]. LOF pores or air pores will preferentially become the corrosion source of pitting, and constantly expand to the interior of the specimen to form corrosion pits. The corrosive medium will further attack and destroy the protective passivation film on the surface. Pore defects significantly affect the stability of the surface passivation film of 316L stainless steel [46].

The formation of defects is related to the complex heat transfer pattern and the dynamic characteristics of the melt pool [47, 48]. A large number of LOF pores and air pores are found in the traditional mode. In the traditional mode, heat accumulation affects the melt pool width and makes it grow continuously. The shape of the melt pool is always changing dynamically, which causes insufficient overlap of the melt pool between layers and creates LOF pores. In addition, under the influence of heat accumulation and the increase of melt pool temperature, the rapid vaporization of liquid metal and water vapor leads to depression of the liquid surface, where bubbles and organic matter have no time to escape, resulting in the formation of pore defects. In addition, the increase of the melt pool width enhances the Rayleigh-plateau instability and promotes porosity generation [49]. The defects are mainly located at the bond between the thin-walled part and the base material or at the interlayer metallurgical bond.

Closed-loop control is introduced into laser directed energy deposition to decrease thermal build-up and improve melt pool width uniformity, thereby reducing porosity defects. In the control mode, the traditional deposition mode is used in the early stages. The melt pool width gradually increases due to the melt pool dynamics and thermal effects. Considering the thermal effect of the melt pool and the heat affected zone (HAZ) during deposition, the total laser input (P_{in}) can be defined as $P_{in} = P_{abs} + P_{ref}$, where the P_{abs} represents the energy that absorbs and forms the melt pool, and P_{ref} is considered to be the energy that is reflected off the surface [50]. Even if a constant laser power is set prior to deposition, the actual power will increase slightly from the set value due to the presence of reflected energy from the melt pool surface in actual production. The increase in laser power is influenced by the HAZ, which causes more powder to be melted and increases the melt pool width. In addition, the increase in melt width is also associated with the flow state of the melt pool. The Peclet number (Pe) [51] is commonly used to analyse the effect of powders on melt pool flow, thermal stability, and oscillation. The thermal conductivity of the molten metal, the density of the alloy, the specific heat and the average velocity of the melt pool are all constant due to the identical scanning speed, then the melt pool width increases. In the later stages of deposition, the laser power input is controlled using a PID control algorithm in order to suppress excessive growth in melt pool width. The average value of the melt pool widths during the first 5 layers is used as the melt pool width set-point for the PID controller. By dynamically adjusting the laser power, the growing trend of melt pool width is effectively suppressed.

The closed-loop control process improves the uniformity of the melt pool width and increases the laminar overlap of the melt pool, which further suppresses LOF porosity. In addition, the reduction of heat input slows down the liquid metal vaporization and water vapor collapse, reducing the Rayleigh-plateau instability and further suppressing the pore generation. During the formation of the melt pool, a large amount of oxides and supersaturated gases are generated. If these oxides cannot escape from the melt pool, inclusions and porosity are generated in the deposited layer, further reducing the denseness and mechanical properties. Combined with the industrial CT results, it can be found that porosity is objectively present in both modes. However, pores exceeding 11×10^4 μm^3 are largely eliminated in the control mode, and the overall pore volume is also reduced by 52.5% compared with the traditional mode. Obviously, the pore density is smaller and less in number in the control mode. Moreover, the suppression of heat accumulation can reduce the melt pool temperature to a certain extent, inhibit the diffusion of C elements to the grain boundary, thus avoiding the precipitation of Cr_3C_2 at the grain boundary by the chemistry reaction between Cr and C. More Cr remains in the intergranular, significantly reducing the generation of Cr-depleted areas and reducing the probability of intergranular corrosion [52, 53]. Therefore, the closed-loop control process plays a significant role in reducing the heat accumulation, improving the melt pool width uniformity and reducing the pore defects. The good deposition quality provides conditions for the formation of a continuous and stable passivation film, which significantly improves the corrosion resistance.

6.5 Tensile Property

6.5.1 Engineering Stress-Strain Curves

The tensile properties of specimens with different sampling directions (parallel and perpendicular) were compared. According to the dimensions, two sets of tensile specimens, along parallel and perpendicular to the deposition direction, were respectively prepared using the wire-cut electric discharge machine. Figure 6.12a, b displays schematic diagrams of the tensile specimens taken from the thin-walled parts with the tensile specimen dimensions. The tensile strength and elongation of the two sets of specimens were tested in the conventional and control modes; the plotted stress-strain curves and statistical histograms are displayed in Fig. 6.12c, d. Three parallel specimens were used as the tensile specimens of each group, distinguished by the same color and different line patterns. It is clear that the strength and toughness of the samples in the two orientations were enhanced after control. For

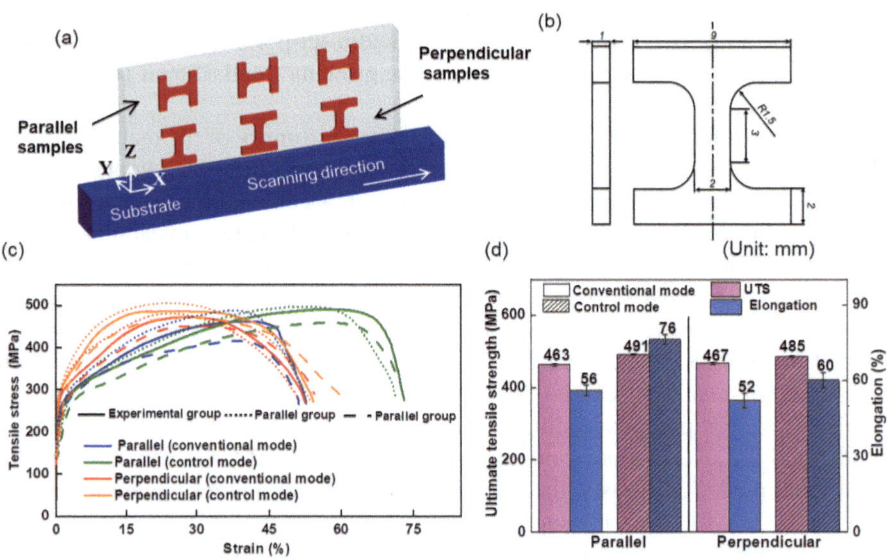

Fig. 6.12 Characterization of tensile properties of parallel and perpendicular specimens in two modes. (**a**) Schematic of parallel and perpendicular samples, named according to the position of the specimen gauge in relation to the scanning direction, (**b**) tensile sample dimensions, (**c**) stress-strain curves of all samples in the conventional and control mode, with different colors representing specimens in different orientations and modes, and the curve line shapes distinguishing the experimental specimens from the parallel groups, and (**d**) histogram of tensile strength and elongation

the parallel specimens, the strength and toughness of the specimens in the control mode were increased by 6.0% and 35.7%, respectively. For the perpendicular specimens, the strength and toughness increased by 3.8% and 15.3%, respectively. Compared with the perpendicular specimens, the enhancement in the tensile properties of the parallel specimens was greater.

6.5.2 *Fractural Morphology*

The fracture morphologies of the parallel specimens are displayed in Fig. 6.13. A shear lip zone with an inclined section was observed at the edge of the fracture surface and a fibrous zone was observed at the fracture. In addition, two pores were observed in the fibrous zone (Fig. 6.13a). When the fibrous zone was magnified 400 times (Fig. 6.13a_1), numerous dimples were observed on the fracture surface, indicating a typical ductile fracture. The dimple morphology was dominated by equiaxed and tearing dimples (Fig. 6.13a_2). Because of the presence of shear stress, the dimples were elliptical under plastic deformation, thereby forming tearing dimples. In the control mode, except for the shear lip and fibrous zones, radial zones were observed on the fracture surface. The unevenness near the fracture was clear necking shrinkage. The radiation zone and necking constriction were the main features of a cup-and-cone fracture and indicated the characteristics of the ductile fracture. In addition, no large readily apparent pores could be observed. By magnifying the

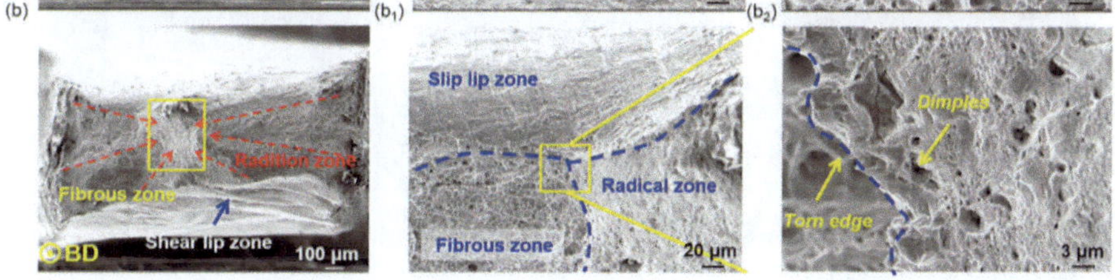

Fig. 6.13 Fracture morphology characterization of parallel tensile specimens. Readily apparent pores are observed in the small magnification fracture (**a**) of the tensile specimen in the conventional mode; the magnification reveals that it is mainly dominated by the dimple network (**a$_1$**) and tearing dimples (**a$_2$**). Conversely, the fibrous and radiation zones can be observed at the small-magnification fracture (**b**) of the tensile specimen in the control mode; typical toughness fracture characteristics such as fracture source traces at the junction of the three zones (**b$_1$**) and dimples and tearing edges (**b$_2$**) can also be observed after magnification

fracture 400 times (Fig. 6.13b$_1$), it could be observed that the shear lip, fibrous, and radiation zones were present simultaneously. The presence of the radiation zone indicates that cracks were generated from the center of the fracture. Further magnification of the intersection of the three zones revealed a tearing edge, representing a clear slip separation fracture. Thus, parallel specimens exhibited ductile fractures with different fracture characteristics. The conventional mode was a shear port and control mode was a cup-and-cone fracture.

The fracture morphologies of the perpendicular samples are displayed in Fig. 6.14. Owing to the different sampling directions, only one deposited layer could be observed in the gauge length of the parallel specimen, whereas multiple layers could be observed in the perpendicular specimen. Therefore, the density of the interlayer bond has a greater influence on the tensile properties of the perpendicular specimen. As indicated in Fig. 6.14a, in addition to the shear lip and fibrous zones, which represent typical ductile fractures, a number of pores were also found. Under tensile loading, the fine pores were easily stretched and formed larger pores. The pores, with a diameter of approximately 15 μm, were found surrounded by a network of equiaxed dimples (Fig. 6.14a$_1$) when magnified to 3000 times. Numerous impurities or second-phase particles were found in the dimple depressions (Fig. 6.14a$_2$), indicating that the inclusions or second-phase particles within the material were susceptible to stress concentration, leading to microscopic pore aggregation and eventual fracture formation. In the closed-loop mode, similar to the parallel specimen, the fracture was characterized by a radiation zone, shear lip zone, and cup-and-cone shape in the fibrous zone. No readily apparent large pores were

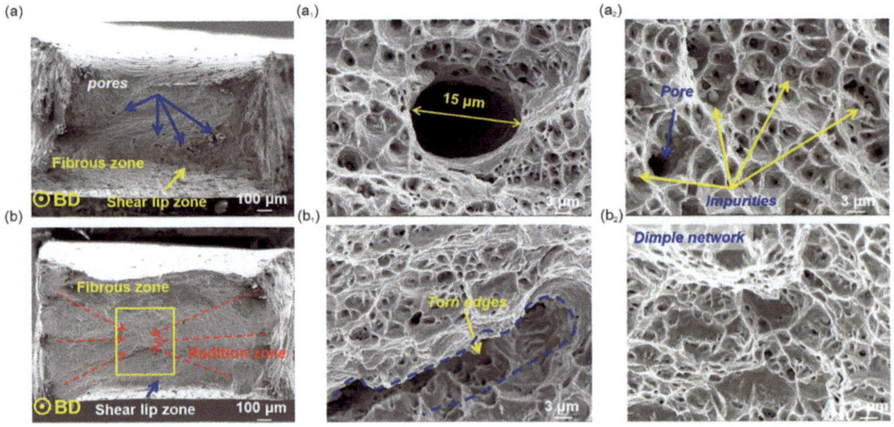

Fig. 6.14 Fracture morphology characterization of perpendicular tensile specimens. A higher number of pores is observed at the small magnification fracture (**a**) of the tensile specimen in the conventional mode, and up to 15 μm in diameter (**a$_1$**). In addition, impurity particles (**a$_2$**) can be observed after magnification. Conversely, virtually no large-diameter pores can be found at the small-magnification fracture (**b**) of the tensile specimen in the control mode, and typical toughness fracture features such as tearing edges (**b$_1$**) and dimple networks (**b$_2$**) can also be observed upon magnification

observed, and the porosity decreased substantially (Fig. 6.14b). The same tearing edges could be observed at the fracture, and the dimples were primarily equiaxed and elliptical (Fig. 6.14b$_1$, b$_2$). It was observed that porosity had a greater influence on the perpendicular specimens. The conventional mode was an aggregated micropore and dimple fracture, whereas the control mode was a slip-separation fracture.

6.5.3 *Fractural EBSD Observation*

Figure 6.15 demonstrates the EBSD results of the parallel tensile specimen fracture in both modes, with the sampling location shown in Fig. 6.15g. As indicated in Fig. 6.15a, regularly distributed band-like structures were observed in the longitudinal section of the sample in the conventional mode. Based on the corresponding kernel average misorientation (KAM) map in Fig. 6.15c, the band-like structures were identified as slip bands with a higher local strain (visualised by higher KAM values, indicated in green). Secondly, a large number of crystal structures were also detected in grains in the <101> || TA direction, distributed at an angle close to 90° to the slip bands. In the control mode, similar regular distributions of slip bands and crystal structures with an approximately 90° distribution were observed (Fig. 6.15b). Compared to the conventional mode with the same sampling area, the difference lies in the fact that the number and width of the slip bands in the control mode were significantly reduced, and the number of crystal structures with a 90° distribution was significantly increased. Figure 6.15d further verifies the distribution characteristics of the slip bands and 90° crystal structures as well as the local strain intensity in the control mode. The highest grain orientation strengths were observed in the <322> direction in the inverse pole figure (IPFs) of both modes, with the difference being that the strength in the control mode was nearly twice as high as that in the conventional mode.

Figure 6.16 illustrates the distribution of grain-boundary orientation angles from the sample of Fig. 6.14, where grain boundaries less than than 10° and greater than 10° are respectively labeled in red and black, and 60° grain boundaries are labeled in blue, which proves that the crystal structures with angular distributions were deformation twins. Based on the orientation diagram in Fig. 6.15a it could be observed that no deformation twins were formed in the <001> and <111> oriented matrix parallel to the TA. A certain number of deformation twins were detected in the grains in the <101> || TA direction. Based on Fig. 6.16c, the twin boundaries exhibited a peak height of 60° with a frequency of 55.8%. In the control mode (Fig. 6.16d), deformation twins were also detected in the grains in the <111> || TA direction with a problability of 64% at 60°. This indicated a significant increase in the number of deformation twins and a decrease in the number of dislocation slips after in-situ control of the melt pool.

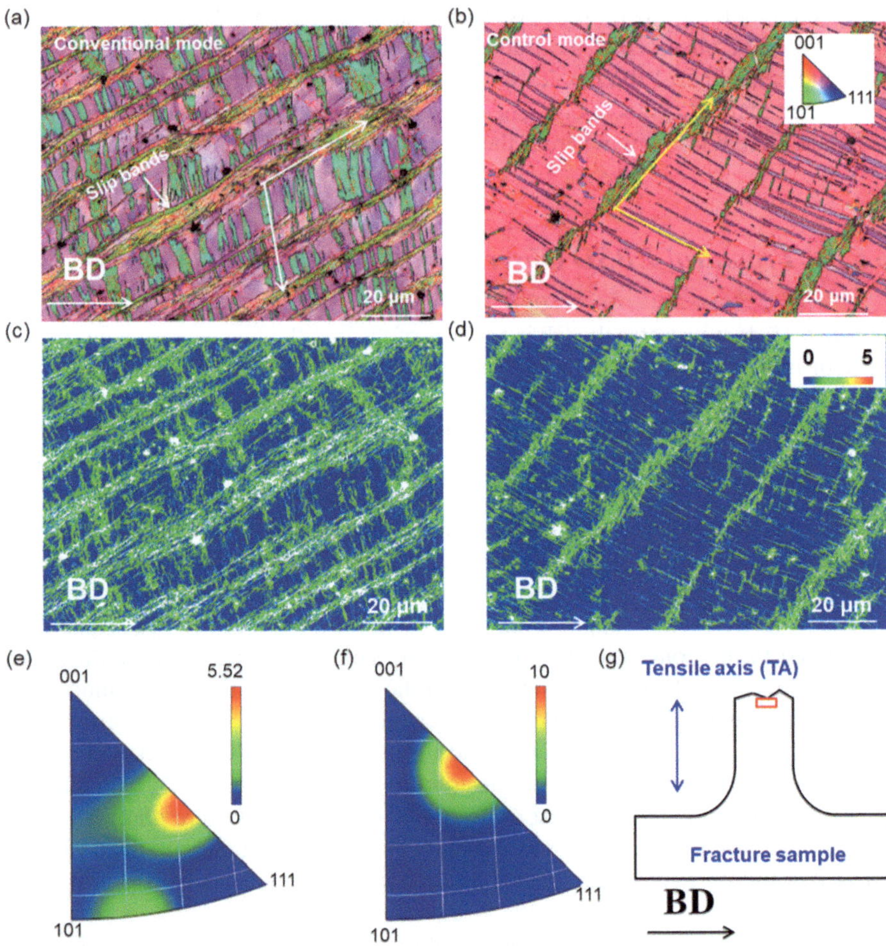

Fig. 6.15 EBSD analysis of samples after fracture. (**a**) IPF orientation map in conventional mode, (**b**) IPF orientation map in control mode, (**c**) corresponding kernel average misorientation map in the conventional mode, (**d**) corresponding kernel average misorientation map in control mode, (**e**) the inverse pole figure along the tensile axis (TA) in conventional mode, (**f**) IPF along TA in control mode and (**g**) sampling position of the EBSD sample

6.5.4 Fractural Mechanism Analysis

By analyzing the fracture morphology of the tensile specimens and EBSD characterization of the longitudinal sections, it was found that the factors influencing the tensile properties of the materials were mainly the quality of interlayer bonding and microstructural differences. Porosity influences the interlayer bond strength, which in turn influences the tensile properties of the specimens [54]. In the conventional mode, the parallel specimens mainly exhibited shear fractures dominated by shear lips and fibrous zones. In addition to the equiaxed dimples, others formed as an

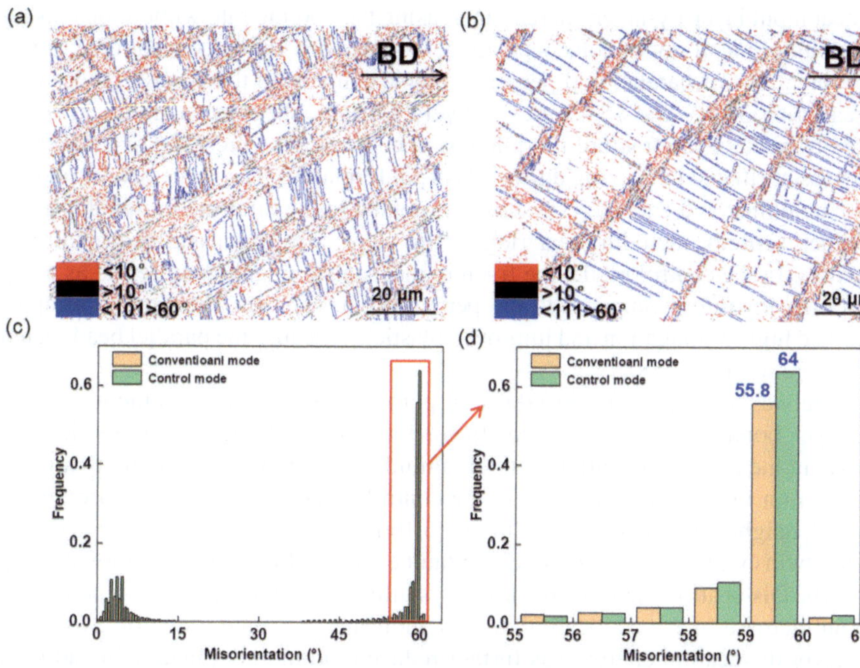

Fig. 6.16 The distribution of different angular grain boundaries in two modes. Distribution of twinned grain boundaries with consistent orientation can be seen in the conventional mode (**a**), and the content of twinned grain boundaries at 60° is significantly increased in the conventional mode (**b**). (**c**) The distribution of grain boundaries at different angles in two modes, and (**d**) the twinned grain boundaries at 60° in the control mode are increased by 14.7% compared with that in the conventional mode

ellipsoidal shape under shear stress, which are called tearing dimple. Two independent pores were observed at the parallel samples fracture, whereas the pores of the perpendicular samples were more densely distributed and smaller. Because the perpendicular specimens were more susceptible to interlayer metallurgical bonding due to the gauge length was perpendicular to the scanning direction. In addition, the continuous high-energy laser resulted in powder spattering, which was vulnerable to inclusions during deposition. As plastic deformation of the material occurred, the inclusions within the material became prone to fracture or separation from the matrix due to stress concentration [55, 56]. The small pores formed during the metallurgical process were further stretched under tensile loading to form large pores. Impurities were observed in the middle of depressions in the dimple network, confirming that the fracture mechanism was an aggregate of micro-pores and dimple fractures. The slow cooling rate and continuous laser deposition in the conventional mode increased the complexity of the thermal cycle and melt pool state as well as the tendency for metallurgical defects [57].

This problem was addressed by changing the laser power and controlling the melt pool width, which reduced the thermal accumulation effect by directly varying the heat input layer-by-layer, thereby increasing the cooling rate without increasing the cooling time or continuous deposition. Compared with the conventional mode, the elongations of parallel and perpendicular specimens in the control mode were increased by 35.7% and 15.4%, respectively. Fibrous, shear lip, and radical zones were observed in the fracture morphology specimens. The presence of the radical zone was attributed to necking during plastic deformation. The successive evolution of the columnar and equiaxed crystals contributed to the different morphologies of the crack sources, including tear ridges, slip steps, and larger and deeper dimple networks. In the control mode, the fracture mechanism mainly manifested as a slip separation fracture. Compared to the perpendicular samples, the parallel samples exhibited higher elongation and improved plasticity because the parallel beads acted as reinforcing fibers [58].

There was no significant difference in porosity between the two modes, so the interlayer bonding was not a major factor influencing the tensile properties. By analyzing the grain distribution in the longitudinal section, it was found that more deformation twins were produced in the control mode, which can be observed in the peak height comparison results in Fig. 6.16d. The increase or decrease in deformation twins was related to the stacking fault energy (SFE) and dislocation density [59, 60]. Dislocations tend to form in crystalline materials at high temperatures, and an increase in the dislocation density indicates the suppression of cross-slip. The layer dislocation energy was further reduced, promoting planar slip and twin generation [61]. The in-situ melt pool control process further promoted more deformation twins by decreasing the dislocation density and SFE through the reduction of thermal accumulation effects. Twin boundaries act as barriers to dislocation motion, effectively reducing the mean free path of dislocations and creating a "dynamic Hall–Petch" effect [62, 63]. This can increase the strain-hardening capacity of the material during deformation, leading to a twinning-induced plasticity (TWIP) effect, which is also the main reason for the elongation increase of the specimen.

The tensile specimen strength and toughness were enhanced simultaneously, and the UTS (ultimate tensile strength) and elongation of tensile specimen parallel to the deposition direction were increased by 6% and 35.7%, respectively. In general, there are two methods to enhance the tensile strength: increasing the dislocation density and reducing the grain size [64, 65]. During the deposition of the thin-walled parts, owing to the specificity of the deposition path, the overall grains were primarily large columnar dendrites that grew across the melt pool boundary. The in-situ melt pool control process increased the cooling and solidification rates of the melt pool by mitigating the thermal accumulation effect, and the interlayer grain size was significantly reduced. In this case, the dislocation density and grain size decreased simultaneously in the control mode; therefore, the improvement in the material tensile strength was not obvious. Overall, the plasticity of the material was significantly improved while ensuring the tensile strength via the in-situ melt pool control.

6.6 Summary

In this chapter, the corrosion behaviour and tensile property of 316LSS thin-walled parts in the traditional and control modes were investigated. The effects of closed-loop control of melt pool width on the roughness, pore defects and electrochemical corrosion resistance were the focus of attention. The reliability of the in-situ melt pool control for improving the surface quality and forming quality of the parts was confirmed by characterizing the dimensional accuracy, surface roughness, and microstructure and pore distribution of the parts. Some key findings were listed below:

1. Affected by thermal accumulation and uneven melt pool shape, LOF pores and porosity defects were easily generated in directed energy deposition, which would preferentially contact with the corrosion medium for pitting corrosion.
2. The closed-loop control process ensured uniformity of melt pool width by reducing laser power and mitigating thermal accumulation effects. The overall reduction of pore volume by 52.5% in the control mode increased the integrity and stability of the passivation film.
3. Potentiodynamic polarization results showed a ~37.6% decrease in passive current density (I_{pass}) and ~18.6% increase in pitting potential (E_{pit}) of the specimens in the control mode.
4. The closed-loop control of the heat input limited the dissolution of C elements to the grain boundaries and the generation of Cr_3C_2, thus retaining more Cr elements between the grains and avoiding excessive intergranular corrosion. The corrosion morphology characterization further confirmed the role of the closed-loop control process in suppressing intergranular corrosion.
5. The in-situ melt pool control reduced the dislocation density and stacking fault energy by reducing the thermal accumulation effect, which further promoted the creation of more deformation twins. This increased the strain-hardening capacity of the material during deformation, producing a twin-induced plasticity (TWIP) effect. The elongation of the samples increased to 35.7%.

References

1. J. Liu, S. Suslov, Z.C. Ren, Y.L. Dong, C. Ye, Microstructure evolution in Ti64 subjected to laser-assisted ultrasonic nanocrystal surface modifcation, Int. J. Mach. Tool Manuf. 136 (2019) 19–33, https://doi.org/10.1016/j.ijmachtools.2018.09.005.
2. Y.K. Zhang, J.Z. Lu, X.D. Ren, H.B. Yao, H.X. Yao, Effect of laser shock processing on the mechanical properties and fatigue lives of the turbojet engine blades manufactured by LY2 aluminum alloy, Mater. Des. 30 (2009) 1697–1703.
3. M. Shakil, M. Ahmad, N.H. Tariq, B.A. Hasan, J.I. Akhter, E. Ahmed, M. Mehmood, M.A. Choudhry, M. Iqbal, Microstructure and hardness studies of electron beam welded Inconel 625 and stainless steel 304L, Vacuum 110 (2014) 121–126, https://doi.org/10.1016/j.vacuum.2014.08.016.

4. D.L. Bourell, M. Leu, D. Rosen, Roadmap for additive manufacturing-Identifying the future of freeform processing. The University of Texas at Austin, Laboratory for Freeform Fabrication, Adv. Manuf. Cent. (2009) 32.
5. B. Bax, R. Rajput, R. Kellet, M. Reisacher, Systematic evaluation of process parameter maps for laser cladding and directed energy deposition, Addit. Manuf. 21 (2018) 487–494, https://doi.org/10.1016/j.addma.2018.04.002.
6. F. Kaji, H. Nguyen-Huu, A. Budhwani, J. A. Narayanan, M. Zimny, E. Toyserkani, A deep-learning-based in-situ surface anomaly detection methodology for laser directed energy deposition via powder feeding, J. Manuf. Process. 81 (2022) 624-637, https://doi.org/10.1016/j.jmapro.2022.06.046.
7. B. Li, Y. Zhang, Y.C. Lei, H.Y. Wei, C. Chen, F. Liu, P.H. Zhao, K.Y. Wang, A single-sensor multi-scale quality monitoring methodology for laser-directed energy deposition: Example with height instability and porosity monitoring in additive manufacturing of ceramic thin-walled parts, Addit. Manuf. 79 (2024) 103923, https://doi.org/10.1016/j.addma.2023.103923.
8. J.J. Han, J. Zhu, W. Zheng, G.C. Wang, Influence of metal forming parameters on surface roughness and establishment of surface roughness prediction model, Int. J. Mech. Sci. 163 (2019) 105093, https://doi.org/10.1016/j.ijmecsci.2019.105093.
9. J.C. de Lacerda, G.D. Martins, V.T. Signoretti, R.L.P. Teixeira, Evolution of the surface roughness of a low carbon steel subjected to fatigue, Int. J. Fatig. 102 (2017) 143-148, https://doi.org/10.1016/j.ijfatigue.2017.05.010.
10. A. Heralica, A. Christianssona, B. Lennartson, Increased stability in laser metal wire deposition through feedback from optical measurements, Opt. Lasers Eng. 48 (2010) 478-485, https://doi.org/10.1016/j.optlaseng.2009.08.012.
11. DebRoy T, Wei HL, Zuback JS, Mukherjee T, Elmer JW, Milewski JO. Additive manufacturing of metallic components – Process, structure and properties. Prog Mater Sci. 92 (2018) 112–224. https://doi.org/10.1016/j.pmatsci.2017.10.001.
12. S. M. Thompson, L.K. Bian, N. Shamsaei, A. Yadollahi, An overview of Direct Laser Deposition for additive manufacturing; Part II: Mechanical behavior, process parameter optimization and control, Addit. Manuf. 8 (2015) 12–35, https://doi.org/10.1016/j.addma.2015.07.002.
13. A. Dash, J. D. Avila, S. Bose, A. Bandyopadhyay, Influence of active cooling on microstructure and mechanical properties of wire arc additively manufactured mild steel, Front. Mech. Eng 9:1130407 (2023), https://doi.org/10.3389/fmech.2023.1130407.
14. D.Q. Yang, G. Wang, G.J. Zhang, Thermal analysis for single-pass multi-layer GMAW based additive manufacturing using infrared thermography, J. Mater. Process. Technol. 244 (2017) 215–224, https://doi.org/10.1016/j.jmatprotec.2017.01.024.
15. S. Zheng, C. Li, Y. Qi, L. Chen, C. Chen, Mechanism of (Mg,Al,Ca)-oxide inclusioninduced pitting corrosion in 316L stainless steel exposed to sulphur environments containing chloride ion, Corros. Sci. 67 (2013) 20–31.
16. A.J. Sedriks, Corrosion of Stainless Steels, Wiley, 1996.
17. R.M. Kain, A.H. Tuthill, E.C. Hoxie, The resistance of types 304 and 316 stainless steels to crevice corrosion in natural waters, J. Mater. Energy Syst. 5 (1984) 205–211.
18. R.F. Schaller, A. Mishra, J.M. Rodelas, J.M. Taylor, E.J. Schindelholz, The role of microstructure and surface finish on the corrosion of selective laser melted 304L, J. Electrochem. Soc. 165 (2018) C234–C242.
19. M. A. Melia, H. A. Nguyen, J. M. Rodelas, E. J. Schindelholz, Corrosion properties of 304L stainless steel made by directed energy deposition additive manufacturing, Corros. Sci. 152 (2019) 20–30, https://doi.org/10.1016/j.corsci.2019.02.029.
20. H.J. Lee, P.L. Narayana, J.H. Kim, C.H. Park, J.K. Hong, J.T. Yeom, T. Lee, S.W. Lee, Effect of interlayer cooling on the microstructure and mechanical properties of titanium alloys fabricated using directed energy deposition, J. Alloy. Compd. 953 (2023) 170202, https://doi.org/10.1016/j.jallcom.2023.170202.
21. Z.J. Qiu, Z.Y. Wang, A.A. Gazder, S.V. Duin, A. Studer, U. Garbe, Q.F. Gu, B.T. Wu, H.L. Zhu, D. Wexler, H.J. Li, Stabilised mechanical properties in Ni-based Hastelloy C276 alloy by addi-

tive manufacturing under different heat inputs incorporated with active interlayer temperature control, Mater. Sci. Eng. A 862 (2023) 144434, https://doi.org/10.1016/j.msea.2022.144434.
22. X. Xu, H.F. Lu, Y.Y. Su, M.X Peng, F. Xing, K.Y. Luo, J.Z. Lu, Comparing corrosion behavior of additively manufactured Cr-rich stainlesssteel coating between conventional and extreme high-speed laser metal deposition, Corros. Sci. 195 (2022) 109976, https://doi.org/10.1016/j.corsci.2021.109976.
23. M. H. Farshidianfar, A. Khajepour, A. P. Gerlich, Real-time control of microstructure in laser additive manufacturing, Int. J. Adv. Manuf. Technol. 82 (2016) 1173–1186.
24. M. H. Farshidianfar, A. Khajepour, A. P. Gerlich, Effect of real-time cooling rate on microstructure in Laser Additive Manufacturing, J. Mater. Process. Technol. 231 (2016) 468-478.
25. Y.Y. Ding, J. Warton, R. Kovacevic, Development of sensing and control system for robotized laser-based direct metal addition system, Addit. Manuf. 10 (2016) 24-35, https://doi.org/10.1016/j.addma.2016.01.002.
26. S. Tallon, C.E. Davies, Use of the attenuation of acoustic pulsed waves for concentration measurement in gas–solid pipe flow. Advances in fluidization and fluid particle systems, AIChE Symp. Ser. 93317 (1997) 136–140.
27. P.S. Cooka, A.B. Murphy, Simulation of melt pool behaviour during additive manufacturing: Underlying physics and progress, Addit. Manuf. 31 (2020) 100909, https://doi.org/10.1016/j.addma.2019.100909.
28. A. Vasinonta, J. Beuth, M. Griffith, Process maps for controlling residual stress and melt pool size in laser-based SFF processes, Proceedings of the Solid Freeform Fabrication Symposium (2000) 200–208.
29. A. Heralica, A. Christianssona, B. Lennartson, Height control of laser metal-wire deposition based on iterative learning control and 3D scanning, Opt. Lasers Eng. 50 (9) (2012) 1230–1241.
30. Y.Y. Su, Z.F. Wang, H.F. Lu, K.Y. Luo, J.Z. Lu, Improved wear resistance of directed energy deposited Fe-Ni-Cr alloy via closed-loop controlling laser power, J. Manuf. Process. 75 (2022) 802-813, https://doi.org/10.1016/j.jmapro.2022.01.047.
31. M. Gharbi, P. Peyre, C. Gorny, M. Carin, S. Morville, P. Le Masson, D. Carron, R. Fabbro, Influence of various process conditions on surface finishes induced by the direct metal deposition laser technique on a Ti–6Al–4V alloy, J. Mater. Process. Technol. 213 (2013) 791–800, https://doi.org/10.1016/j.jmatprotec.2012.11.015.
32. H. C R, R. J R. Mechanism for Minor Element Effect on GTA Fusion Zone Geometry. Welding Journal, 1982, 61(4): 97-102.
33. M.G. Pujar, U.K. Mudali, S.S. Singh, Electrochemical noise studies of the effect of nitrogen on pitting corrosion resistance of high nitrogen austenitic stainless steels, Corros. Sci. 53 (2011) 4178–4186.
34. Y. Qiu, S. Thomas, M.A. Gibson, H.L. Fraser, N. Birbilis, Corrosion of high entropy alloys, Npj Mater. Degrad. 1 (2017) 15, https://doi.org/10.1038/s41529-017-0009-y.
35. A.M. Ribeiro, A.C. Alves, L.A. Rocha, F.S. Silva, F. Toptan, Synergism between corrosion and wear on CoCrMo−Al2O3 biocomposites in a physiological solution, Tribol. Int. 91 (2015) 198–205.
36. I. Sulima, R. Kowalik, P. Hyjek, The corrosion and mechanical properties of spark plasma sintered composites reinforced with titanium diboride, J. Alloys. Compd. 688 (2016) 1195–1205.
37. S. Bagherifard, S. Slawik, I. Fernández-Pariente, C. Pauly, F. Mücklich, M. Guagliano, Nanoscale surface modifcation of AISI 316L stainless steel by severe shot peening, Mater. Des. 102 (2016) 68–77.
38. D. Kirk, N.J. Payne, Transformations induced in austenitic stainless steel by shot peening, Presented at the The Seventh International Conference on Shot Peening (1999).
39. Y. Bai, S.J. Li, F. Prima, Y.L. Hao, R. Yang, Electrochemical corrosion behavior of Ti–24Nb–4Zr–8Sn alloy in a simulated physiological environment, Appl. Surf. Sci. 258 (8) (2012) 4035–4040.
40. I. Milošev, G. Žerjav, J.M. Calderon Moreno, M. Popa, Electrochemical properties, chemical composition and thickness of passive flm formed on novel Ti–20Nb–10Zr–5Ta alloy, Electrochim. Acta 99 (2013) 176–189.

41. M.C. Biesinger, B.P. Payne, A.P. Grosvenor, L.W.M. Lau, A.R. Gerson, R.S.C. Smart, Resolving surface chemical states in XPS analysis of frst row transition metals, oxides and hydroxides: Cr, Mn, Fe, Co and Ni, Appl. Surf. Sci. 257 (2011) 2717–2730, https://doi.org/10.1016/j.apsusc.2010.10.051.
42. A.R. Brooks, C.R. Clayton, K. Doss, Y.C. Lu, On the Role of Cr in the Passivity of Stainless Steel J. Electrochem. Soc. 133 (1986) 2459–2464. https://iopscience.iop.org/article/10.1149/1.2108450
43. C.O.A. Olsson, D. Landolt, Passive flms on stainless steels—chemistry, structure and growth, Electrochim. Acta 48 (2003) 1093–1104, https://doi.org/10.1016/S0013-4686(02)00841-1.
44. D.C. Kong, C.F. Dong, X.Q. Ni, X.G. Li, Corrosion of metallic materials fabricated by selective laser melting. npj Mater. Degrad. 3 (2019) 24, https://doi.org/10.1038/s41529-019-0086-1.
45. C.Y. Wang, K.Y. Luo, X.Y. Bu, Y.Y. Su, J. Cai, Q.L. Zhang, J.Z. Lu, Laser shock peening-induced surface gradient stress distribution and extension mechanism in corrosion fatigue life of AISI 420 stainless steel, Corro. Sci. 177 (2020), 109027, https://doi.org/10.1016/j.corsci.2020.109027.
46. D.C. Kong, C.F. Dong, X.Q. Ni, L. Zhang, H. Luo, R.X. Li, L. Wang, C. Man, X.G. Li, The passivity of selective laser melted 316L stainless steel, Appl. Surf. Sci. 504 (2020), 144495, https://doi.org/10.1016/j.apsusc.2019.144495.
47. J. Reijonen, A. Revuelta, T. Riipinen, K. Ruusuvuori, P. Puukko, On the effect of shielding gas flow on porosity and melt pool geometry in laser powder bed fusion additive manufacturing, Addit. Manuf. 32 (2020) 101030, https://doi.org/10.1016/j.addma.2019.101030.
48. B. Lanea, I. Zhirnov, S. Mekhontsev, S. Grantham, R. Ricker, S. Rauniyar, K. Chou, Transient Laser Energy Absorption, Co-axial Melt Pool Monitoring, and Relationship to Melt Pool Morphology, Addit. Manuf. 36 (2020) 101504, https://doi.org/10.1016/j.addma.2020.101504.
49. I. Yadroitsev, A. Gusarov, I. Yadroitsava, I. Smurov, Single track formation in selective laser melting of metal powders, J. Mater. Process. Technol. 210 (2010) 1624-1631, https://doi.org/10.1016/j.jmatprotec.2010.05.010.
50. X.H. Zhou, D.L. Li, L. Zhang, Q.L. Duan, Application of an adaptive PID controller enhanced by a differential evolution algorithm for precise control of dissolved oxygen in recirculating aquaculture systems, Biosyst. Eng. 208 (2021) 186-198, https://doi.org/10.1016/j.biosystemseng.2021.05.019.
51. R.J. Williamsa, A. Piglioneb, T. Rønneberga, C. Jonesc, M. Phamb, C.M. Daviesa, P.A. Hooper, In situ thermography for laser powder bed fusion: Effects of layer temperature on porosity, microstructure and mechanical properties, Addit. Manuf. 30 (2019) 100880, https://doi.org/10.1016/j.addma.2019.100880.
52. S. Gollapudi, Grain size distribution effects on the corrosion behaviour of materials, Corros. Sci. 62 (2012) 90–94, https://doi.org/10.1016/j.corsci.2012.04.040.
53. L. Jinlong, L. Hongyun, Comparison of corrosion properties of passive flms formed on phase reversion induced nano/ultrafne-grained 321 stainless steel, Appl. Surf. Sci. 280 (2013) 124–131, https://doi.org/10.1016/j.apsusc.2013.04.108.
54. G. Sun, R. Zhou, J. Lu, J. Mazumder, Evaluation of defect density, microstructure, residual stress, elastic modulus, hardness and strength of laser-deposited AISI 4340 steel, Acta Mater. 84 (2015) 172–189.
55. S.H. Li, D.H. Lee, Y. Zhao, U. Ramamurty, Hydrogen-induced softening and embrittlement in 316L stainless steel fabricated using laser-powder bed fusion, Acta Mater. 274 (2024) 119959.
56. Y.S. Liu, H.R. Chen, R. Han, Y. Peng, K.H. Wang, X.P. Li, Investigating the microstructure and mechanical properties of 316L/TiB2 composites fabricated by laser cladding additive manufacturing, J. Mater. Res. Technol. 29 (2024) 28-39.
57. L. Wang, P. Pratt, S.D. Felicelli, H. El Kadiri, J.T. Berry, P.T. Wang, M.F. Horstemeyer, Experimental analysis of porosity formation in laser-assisted powder deposition process, TMS Annual Meeting & Exhibition, (2009).
58. P. Guo, B. Zou, C. Huang, H. Gao, Study on microstructure, mechanical properties and machinability of efficiently additive manufactured AISI 316L stainless steel by high-power direct laser deposition, J. Mater. Process. Technol. 240 (2017) 12-22.

59. D. Molnár, X. Sun, S. Lua, W. Lia, G. Engberg, L. Vitos, Effect of temperature on the stacking fault energy and deformation behaviour in 316L austenitic stainless steel, Mater. Sci. Eng. A 759 (2019) 490-497, https://doi.org/10.1016/j.msea.2019.05.079.
60. P. Chowdhury, H. Sehitoglu, W. Abuzaid, H. Maier, Mechanical response of low stacking fault energy co-ni alloys-continuum, mesoscopic and atomic level treatments, Int. J. Plast. 71 (2015) 32-61.
61. B.K. Zhao, P. Huang, L.B. Zhang, S.Z. Li, Z. Zhang, Q. Yu, Temperature Efect on Stacking Fault Energy and Deformation Mechanisms in Titanium and Titanium-aluminium Alloy, Sci. Rep. (2020) 10:3086, https://doi.org/10.1038/s41598-020-60013-6.
62. H. Beladi, I.B. Timokhina, Y. Estrin, J. Kim, B.C. De Cooman, S.K. Kim, Orientation dependence of twinning and strain hardening behaviour of a high manganese twinning induced plasticity steel with polycrystalline structure, Acta Mater. 59 (20) (2011) 7787–7799.
63. J. Chen, H.Y. Wei, K. Bao, X.F. Zhang, Y. Cao, Y. Peng, J. Kong, K.H. Wang, Dynamic mechanical properties of 316L stainless steel fabricated by an additive manufacturing process, J. Mater. Res. Technol. 11 (2021) 170-179.
64. Y. Balit, E. Charkaluk, A. Constantinescu, Digital image correlation for microstructural analysis of deformation pattern in additively manufactured 316L thin walls, Addit. Manuf. 31 (2020) 100862, https://doi.org/10.1016/j.addma.2019.100862.
65. B. Barkia , P. Aubry, P. Haghi-Ashtiani, T. Auger, L. Gosmain, F. Schuster, H. Maskrot, On the origin of the high tensile strength and ductility of additively manufactured 316L stainless steel: Multiscale investigation, J. Mater. Sci. Technol. 41 (2020) 209-218, https://doi.org/10.1016/j.jmst.2019.09.017.

Chapter 7
Effect of the Melt Pool In-Situ Control on the Microstructure of Laser Directed Deposition 316L Stainless Steel Curved Thin-Walled Parts

7.1 Introduction

The constantly pursuing progress in metal material processing technology in terms of forming accuracy, energy saving, productivity, and mechanical properties has spawned reforms in metal material processing technology. The recently emerged laser directed energy deposition (LDED) technology (powder based in this work) uses layer-by-layer deposition of materials to prepare prototype parts [1–3]. With the advantages of high geometric flexibility and complexity, low material waste and high manufacturing resolution, LDED has been successfully applied to repair and remanufacture titanium alloys [4], stainless steels [5], nickel-based alloys [6] or other critical materials. These materials are used in a wide range of applications such as in the aerospace and automotive industries, including housings, radiators, and manifolds [7, 8]. These parts often have complex geometrical features such as planar thin-walls, lattice/honeycomb structures, curved-surfaces, and so on.

Due to the special deposition structure and complex curvature, LDEDed curved surface thin-walled components face serious thermal gradient problems due to the large amount of heat input. As the number of deposited layers accumulates, the thermal effect intensifies. Yang et al. [9] analyzed the causes of problems such as collapse and surface deformation when deposited tilted thin-walled parts such as blades. He changed the tilting angle between the fusion head and the substrate to improve the yield strength and ultimate tensile strength of the components. A thermal analysis method suitable for depositing large horizontal parts to simulate temperature and phase transitions was presented by Patrault [10]. The results showed a good agreement between the model and the experimental results with an error of only 2.6%. Silveira et al. [11] investigated the relationship between the cooling rate in intrinsic heat treatment (IHT) and the phase transformation in martensitic steels, confirming that the thermal accumulation effect affects the dislocation density in austenitic stainless steels. If more measures are not taken to slow down the heat

K. Luo et al., *Melt Pool In-Situ Controlling of Laser Additive Manufacturing Technology*, https://doi.org/10.1007/978-981-95-4597-1_7

accumulation, the continuous LDED deposition process will lead to an increase in the substrate temperature and poor heat dissipation conditions in the melt pool. It will also contribute to the increase in the melt pool size along the build-up direction and bring about inhomogeneous trajectory morphology or serious collapse problems [12, 13]. In order to ensure good moulding accuracy and shape of the components, LDED usually requires proper thermal control to reduce heat accumulation.

Currently, common processes to mitigate the heat build-up effect include real-time adjustment of laser process parameters (power, scanning speed, and powder speed), layer-by-layer reduction of heat input, and increased interlayer cooling time (i.e., increasing the dwell time interval for each layer deposition). Heralic et al. [14, 15] compensated for deviations in layer height by varying the wire feed rate through iterative learning control. Hwang et al. [16] proposed an optimal interlayer cooling (IC) step based on the intrinsic equation to ensure the geometrical stability of LDEDed 316L SS within a short period of time. Debroy et al. [17] obtained well-formed Ti6Al4V thin-walled structures by adjusting the heat input to each droplet during pulsed plasma arc welding to mitigate heat build-up. Duan et al. [18] used pulsed voltage to control the alternate ignition and extinction of the plasma arc during the additive manufacturing process. This increased the proportion of discharge energy allocated to the filler wire, thus reducing the overall heat input required for material deposition, which resulted in a reduction by 37% when compared to the conventional approach. Nevertheless, quantitatively analysing the heat accumulation and consumption during LDED is difficult, consequently, the directly controlling of the heat input is limited.

In the LDED process, the melt pool, as a carrier for propagating heat, is heated to form different shapes of oval with the stacking of layers, which affects the surface quality and overall performance of the components [19, 20]. Compared with the traditional methods to mitigate the thermal effect, the melt pool state needs to be measured online and controlled in real time to realise the online feedback of the thermal history and to provide a concrete basis for mitigating the thermal accumulation effect. Jeon et al. [21] developed an online melt pool depth estimation technique for the DED process using a coaxial infrared (IR) camera, a laser line scanner and an artificial neural network (ANN) to estimate the overall accuracy of the melt pool depth. Zheng et al. [22] proposed a method for in-situ thermal management (ISTM)-assisted plasma arc additive manufacturing and investigated the effect of thermal history on PDAS and grain size. The results showed that the ISTM method could significantly alleviate the microscopic segregation and reduce the brittle phase between dendrites by 61.9% compared with the conventional cooling strategy. Lin et al. [23] used the threshold method combined with connected component analysis to extract the melt pool and splatter, and the k-means clustering algorithm to establish the link between the melt pool state and process parameters, providing a basis for quality control. Currently, the monitoring and feedback of thermal history of components are mostly single shapes such as thin-walled parts and blocks. However, the relationship between the thermal history of curved components and the complex curvature is not clear. The moulding accuracy and performance of LDEDed curved-surface thin-walled components have a strong correlation with the thermal history (including the cooling rate, solidification rate, and the number of thermal cycles). To analyse the heat accumulation pattern of curved thin-walled components with different curvatures and to further evaluate the effect of regulating the heat accumulation effect on the microstructure and properties is an urgent problem.

Therefore, this chapter proposed an ISTC process, which took the melt pool width as the control target and used PID control to regulate the laser power in real time to eliminate the thermal accumulation effect. In order to eliminate the steady state error and avoid integral overshoot, the adaptive integral separated PID control algorithm was raised, which further improved the overall computational efficiency and control accuracy. The surface roughness, moulding accuracy and tensile properties of AISI 316L SS curved-surface thin-walled parts deposited by ISTC-assisted LDED were systematically investigated. A mathematical model of the thermal accumulation effect of complex curvature components was established, and the relationship between the thermal effect and the input energy and frequency was deduced. The correspondence between the curvature position and primary dendrite arm spacing (PDAS) was revealed, and the feasibility of the ISTC process to improve the cooling rate and enhance the dimensional accuracy was demonstrated. In addition, the symbiotic relationship between thermal effects and dislocations was discovered, and twin-induced plastic (TWIP) deformation was promoted by mitigating the thermal cumulative effect. The elongations of the controlled specimens successfully reached quasi-superplasticity.

7.2 Experiment and Characterizations

7.2.1 LDED Process

Figure 7.1a shows the main components of the ISTC-assisted LDED equipment, where a CCD camera (Daheng MER2-041-302GM-P camera) is mainly used for melt pool image capture and an infrared thermal imager (Fotric 340+) is used to monitor the melt pool temperature. The average emissivity of deposited 316L stainless steel was calculated to 0.6 by performing a simple blackbody calibration and referenceing previous work [24]. This emissivity was kept constant throughout the whole process and real-time temperature calculations. The CCD camera was captured at 50 fps with a resolution of 720 × 540 pixels. In the deposition process, the image acquisition and processing speed was 1000 fps, so the whole process took 1 ms. In addition, the image acquisition, image processing and analogue signal sending generated a time lag of about 50 ms. In the LDED system, an Etelux circulation purification system was used to control the oxygen content in the deposition process with an accuracy of ±3 ppm oxygen. In addition, it contained molecular sieves and filters, which can filter the shielding gas and dislodge water and dust. In this experiment, the oxygen content in the deposition chamber was set to be less than 100 ppm, and the low oxygen condition ensured the powder to have better fluidity. The scanning paths and deposition strategy for fabricating curved-surface

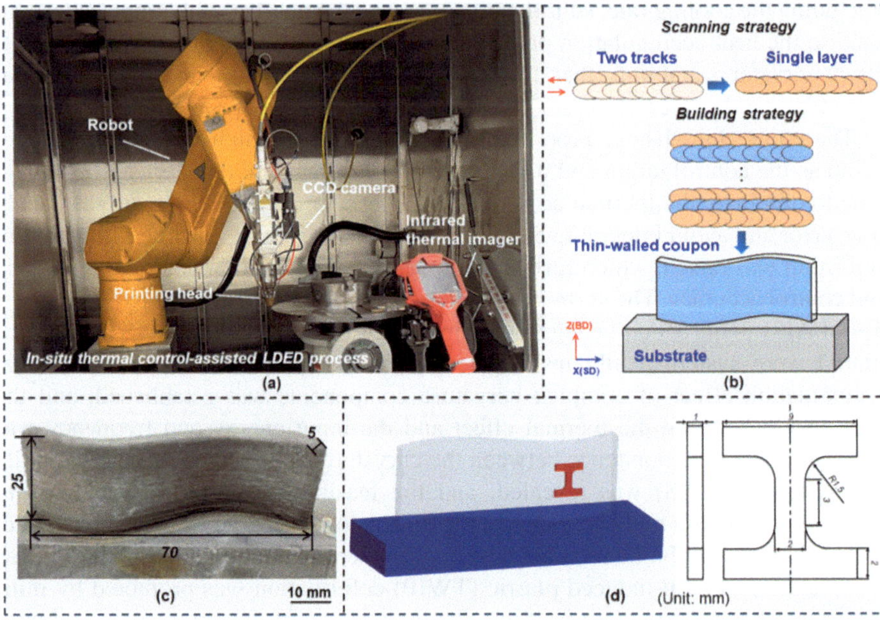

Fig. 7.1 Schematic, experimental setup, and scanning path for in-situ thermal control-assisted laser directed energy deposition processing. (**a**) Equipment components of an in-situ thermal control-assisted laser directed energy deposition process, (**b**) scanning path and building path, (**c**) curved-face thin walled part, (**d**) the sampling location and size of tensile specimens

Table 7.1 Parameters of the deposition in the laser directed energy deposition

Parameters	Value	Units
Powder feed rate	24	g/min
Scanning speed	20	mm/s
Overlapping rate	50%	/
Spot diameter	3	mm

thin-walled parts are shown in Fig. 7.1b. The single layer was formed by reciprocally depositing two tracks in different directions in the X0Y direction, and thin-walled parts were prepared using the same deposition strategy in the Z direction. It was worth noting that the scanning path of the laser head is not a straight line but a sinusoidal curve (the curve equation is described below), which resulted in the formation of curved-surfaces. Since the machine tool used is a five-axis CNC machine tool, the repeat positioning accuracy of a single axis is 0.02 mm. It also has high interpolation accuracy, with an interpolation period of 4 ms; the maximum line speed of the machine tool is 200 mm/s. When taking the plane two-dimensional path, the accuracy of the line speed along the path in the forward direction can be within ±0.5 mm/s. In this study, 45 steel and 316L (powder size: 50–100 μm) were used as the substrate and deposited layers, respectively. The fabricated curved-surface thin-walled parts (70 mm × 5 mm × 25 mm) are shown in Fig. 7.1c. The process parameters are presented in Table 7.1.

7.2.2 Melt Pool Control Strategy

The process design framework for ISTC-assisted LDED is shown in Fig. 7.2. The roles of the CCD camera and infrared thermal imager were melt pool image monitoring and temperature detection, respectively (Fig. 7.2b). The CCD camera was mounted coaxially on the laser head so that the camera could take a real-time image of the top of the melt pool during the deposition process, which also ensured consistency of the position when selecting the melt pool widths of different layers for comparison. The image processing algorithm was used to extract the melt pool width and collect the data to plot the melt pool width curve [25]. By monitoring the temperature in real time, the temperature gradient data can be collected and the cooling rate can be calculated (Fig. 7.2c). In the conventional deposition mode, the detected melt pool widths and temperatures showed the rising tendency. To ensure its stability, an adaptive integral separation PID control was proposed, as shown in Fig. 7.2d.

First, the $w_s(t)$ was set as the set-point value of the ideal melt pool width. There is no specific standard to clarify the ideal value of the melt pool width in the LDED process, so there is no reference basis for the setting of $W_s(t)$. The thermal

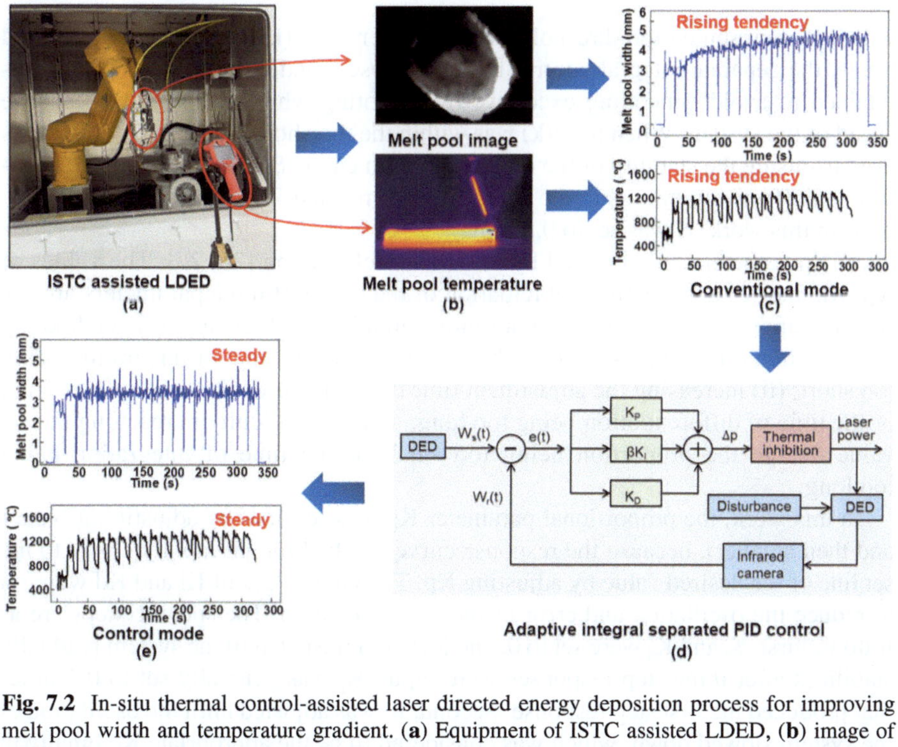

Fig. 7.2 In-situ thermal control-assisted laser directed energy deposition process for improving melt pool width and temperature gradient. (**a**) Equipment of ISTC assisted LDED, (**b**) image of width and temperature of melt pool, (**c**) width and temperature curves of melt pool in the conventional mode, (**d**) adaptive integral separated PID control process, (**e**) width and temperature curves of melt pool in the control mode

accumulation effect is small during the early stage of LDED, so the average width of first three layers of each conventional deposition was taken as $W_s(t)$. The width acquired by the CCD camera was set to the real-time value $w_r(t)$. The difference e(t) between the $w_s(t)$ and $w_r(t)$ was used as an input to the entire ISTC system. The stable melt pool width was obtained by varying the laser power. In order to be recognized and implemented by the processor, the formula was discretized to set the variation of the laser power to

$$\Delta P = K_P\left\{e(k) + \beta\frac{1}{T_I}\sum_{k=1}^{n}e(k) + T_D\left[e(k) - e(k-1)\right]\right\} \tag{7.1}$$

where the PID parameters (K_p, K_i, K_d) determined the sensitivity of this control system. It was worth noting that in PID control, the integral coefficient Ki and the differential coefficient Kd were generally expressed as parameters with respect to the sampling time T: T_I denoted the integral time, and T_D denoted the differential time. The range of values of β satisfied that

$$\beta = \begin{cases} 0, |e(k)| > X \\ 1, |e(k)| \le X \end{cases} \tag{7.2}$$

where X represented the threshold of e(k). When the e(k) exceeded the threshold range, the separate-integral control mode was used, and only the PD control was retained to avoid generating excessive overshooting which affected the response speed of the system. When the e(k) was within the threshold range, PID control was used to ensure the stability of the melt pool width curve. Separate integral PID control can effectively eliminate the steady state error and improve the control accuracy. In this work, X was set to 0.5 mm.

PID parameters are obtained by empirical field tuning [26, 27]. The values of Kp, Ki, and Kd determine the fluctuation of the curve. If three parameters are not selected appropriately, the following three situations will occur: (i) overshooting phenomenon due to the proportion being too big or the time of integration being too short; (ii) increasing the adjustment time due to the proportion being too small or the time of differentiation being too long; and (iii) difficulty in reaching the set value due to the proportion being too small or the time of integration being too long.

In this work, the proportional parameter Kp was set first (by adjusting it bigger and then smaller), because the response curve can be brought rapidly closer to the set line of the desired value by adjusting Kp. Then the values of Ki and Kd were set to reduce the oscillation and error of the response curve. The specific steps are as follows: first, K_i and K_d were set to 0, and K_p was adjusted until the system gradually stabilizes after three step responses were input. K_p was generally set to 0.7 times that produced the first step response. Second, K_i was adjusted until the oscillation of the system slowed down, which was considered to be the appropriate K_i. Finally th e value of K_d was generally 0 or 0.05. In conclusion, the PID parameters needed to be determined by the fluctuation range of the standard melt pool width, which was finally determined as $K_p = 50$, $T_i = 800$, and $T_d = 0.05$.

7.2.3 *Materials Characterization*

In this study, the overall morphology and microstructure of the cross-section of curved-surface thin-walled parts were automatically photographed and stitched using an ultra-depth-of-field microscope (KEYENCE VHX-7000 digital microscope). The fractural morphology of the tensile samples was photographed using a scanning electron microscope (SEM) model S-3400N. The inverse pole figure (IPF) maps and twinnign boundaries of the grains were obtained by an electron backscatter diffraction (EBSD) model Nordly max3. The surface morphology and roughness statistics of the members were measured by a Leica TCS SP5 type laser confocal system.

7.3 Forming Quality Analysis of Original Component

Figure 7.7 shows the curved-surface thin-walled parts deposited with different curvatures (K_1-K_5) and different laser powers (800–2400 W). The curvature is determined by the sinusoidal formula $Y = -a\sin\left(\frac{\pi x}{35}\right)$ $x \in [0, 70]$. Different curvatures were produced when different values of a (unit: mm^{-1}) were taken, and the five curvatures were displayed in Fig. 7.3b. The directions of five curves are the same of the deposition direction of components in Fig. 7.3a. From Fig. 7.3a, it can be seen that when the laser power was 800 W, the surface of the parts showed obvious spheroidisation and viscous powder phenomenon [28]. When increased to 2000 W, the spheroidisation on the surface of the component disappeared and an obvious corrugation appeared. When the initial power was 2400 W, with the rapid increase in the curvatures and the number of layers, the thin-walled parts could not cool down and solidify in time when deposited the next layer, which led to bumps phenomenon on the surface due to gravity [29]. Further analysis (Fig. 7.3c) revealed that the surface roughness of components was less than 20 μm at 1200 W, 1600 W, and 2000 W. However, when the laser powers were 800 W and 2400 W with curvatures of K_3–K_5, the surface roughness were greater than 20 μm, and the highest one reaching 35 μm. When the curvature is K_5, the obvious difference of the surface roughness can be found in samples deposited at different powers. The surface roughness increases by ~48.5% and ~47.5% relative to the minimum value when the power is 800 W and 2400 W, respectively. This implies that the surface roughness could be affected by sticky powder phenomenon and incomplete solidification (Table 7.2).

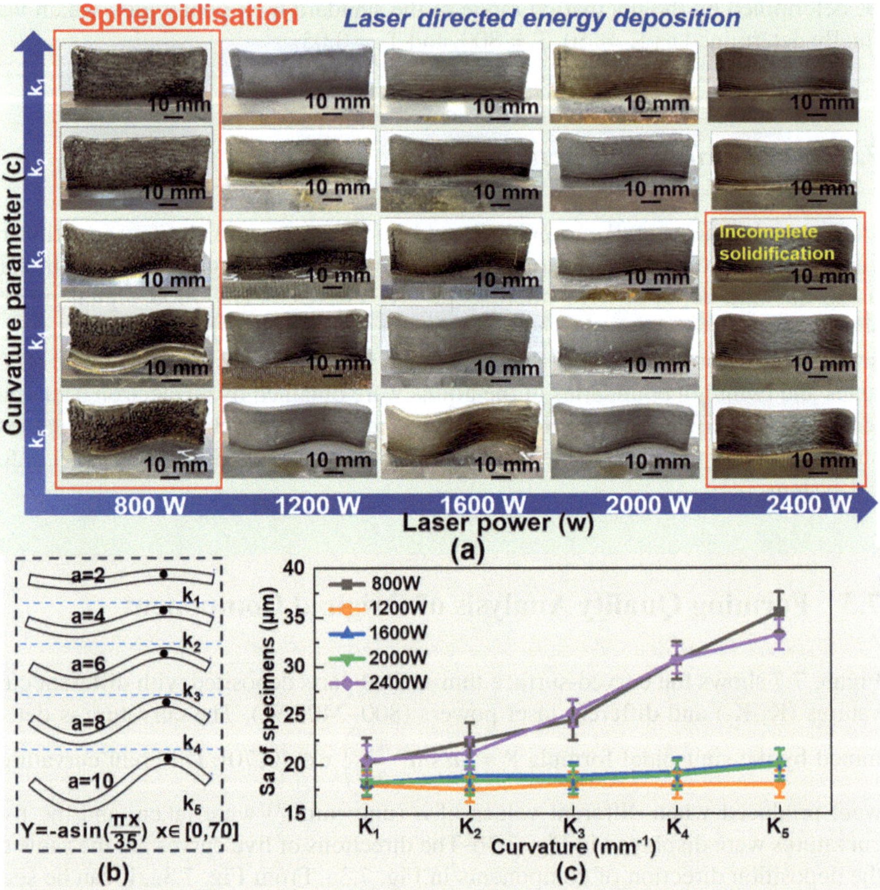

Fig. 7.3 Physical drawing of curved-surface thin-walled parts with roughness results. (**a**) The components deposited at different power and different curvatures, (**b**) five curvatures of deposited specimens, (**c**) roughness results

Table 7.2 Surface roughness of the samples deposited at different laser powers and curvatures

Groups	800 W	1200 W	1600 W	2000 W	2400 W
K_1	20.258 ± 2.534	17.986 ± 1.324	18.208 ± 1.308	18.258 ± 1.258	20.534 ± 1.524
K_2	22.325 ± 1.987	17.542 ± 1.289	18.905 ± 1.419	18.549 ± 1.453	21.208 ± 1.622
K_3	24.585 ± 2.045	18.503 ± 1.265	18.924 ± 1.399	18.545 ± 1.459	25.326 ± 1.389
K_4	30.648 ± 1.899	17.964 ± 1.322	19.325 ± 1.405	19.028 ± 1.489	30.725 ± 1.433
K_5	35.542 ± 2.034	18.028 ± 1.359	20.231 ± 1.392	19.988 ± 1.558	33.252 ± 1.528

7.4 ISTC of Multiple Laser Powers for Manufacturing Curved Components

7.4.1 *Melt Pool Temperature and Melt Pool Images Detection*

Based on the above findings, we focused on the comparison of the ISTC-assisted LDED process for 800 W and 2400 W components. To further validate the universality of the process, 1600 W was also chosen. As shown in Figs. 7.4, 7.5, and 7.6, at the end of depositing different layers, the instantaneous melt pool temperatures and melt pool images for both conventional and control modes are demonstrated. At 800 W (Fig. 7.4a), the instantaneous temperature at the end of the first layer was ~561.5 °C, which increased to ~1848.8 °C at the end of the 15th layer and then decreased to ~1325.7 °C at the 19th layer. There was no further heat transfer at the end of the deposition, so the temperature decreased when the last layer was deposited. Under the ISTC process, the melt pool temperature was effectively reduced by adjusting the heat input, which produced a more pronounced effect as the number of layers increased. For example, in the tenth layer, the instantaneous temperature was ~1631.8 °C, which was reduced by ~148.9 °C. The melt pool image monitored by the CCD camera is shown in Fig. 7.4b, where the melt pool width extracted based on the tangent circle algorithm at the end of the first layer deposited in the

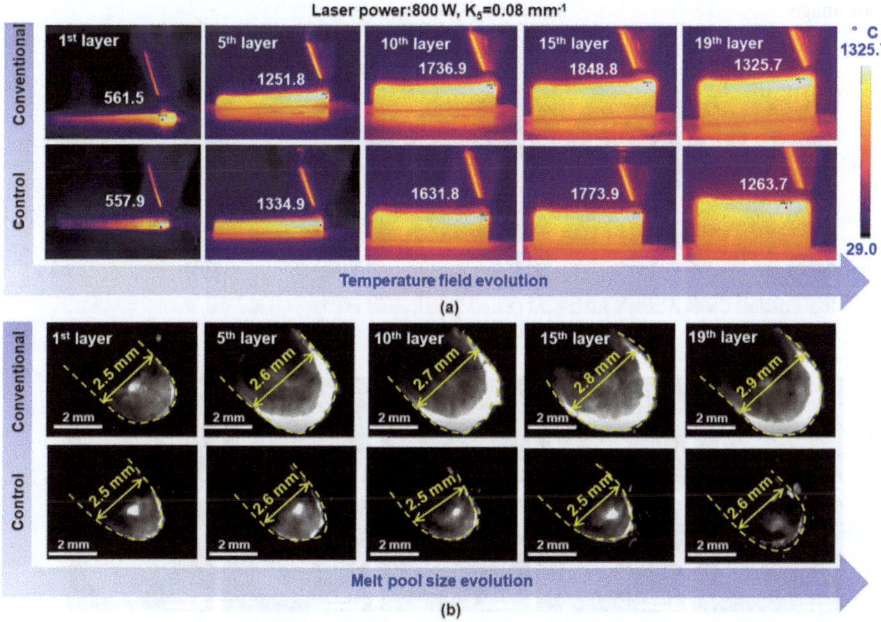

Fig. 7.4 Comparison of melt pool condition at different stages with in-situ thermal control method when the laser power is 800 W and the curvature is 0.08 mm^{-1}. (**a**) Thermal condition, (**b**) melt pool image

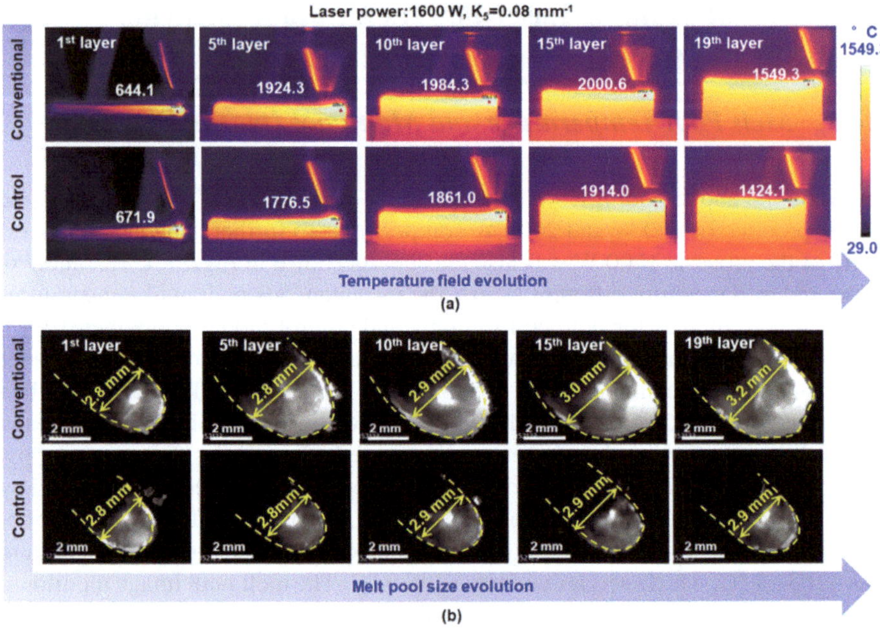

Fig. 7.5 Comparison of melt pool condition at different stages with in-situ thermal control method when the laser power is 1600 W and the curvature is 0.08 mm^{-1}. (**a**) Thermal condition, (**b**) melt pool image

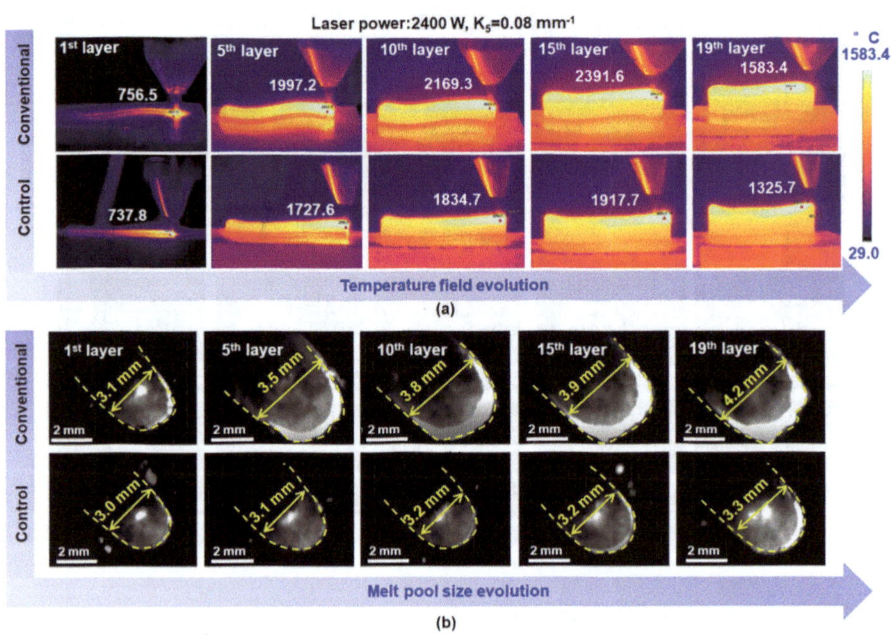

Fig. 7.6 Comparison of melt pool condition at different stages with in-situ thermal control method when the laser power is 2400 W and the curvature is 0.08 mm^{-1}. (**a**) Thermal condition, (**b**) melt pool image

conventional mode was ~2.5 mm. As the number of layers increased and the heat accumulated, the melt pool width increased, and reached ~2.9 mm in the 19th layer. With the introduction of the ISTC method, the stability of the melt pool width was significantly improved. The melt pool widths under the five moments were all in the range of ~2.5 to ~2.6 mm.

From Fig. 7.5a, when the laser power was 1600 W, the instantaneous temperature of the melt pool reached a maximum of ~2000.6 °C, which was an increase of ~151.8 °C compared to the same layer at 800 W. The heat accumulation in the melt pool was mitigated in the control mode, and the melt pool temperature was reduced at each stage. The melt pool width in the conventional mode also increased with the number of deposited layers up to ~3.2 mm. With the introduction of ISTC, the melt pool width can be stabilised at ~2.8–2.9 mm.

When the laser power is 2400 W (Fig. 7.6), the changed rules of melt pool temperature and width in the conventional mode are similar. As the laser power increased, the melt pool temperature increased faster. Obviously, in order to ensure the stability of the melt pool width, the ISTC process has a greater degree of control for the components deposited at higher power. For example, when depositing the 15th layer, the instantaneous melt pool temperature in the controlled mode was 1917.7 °C, which was ~473.9 °C lower compared to the conventional mode. The melt pool width monitored by the camera increased from ~3.1 mm to ~4.2 mm at the end of the deposition of the 19th layer, an increase of ~35.5% in width. In contrast, the melt pool width under ISTC was maintained at ~3.0–3.3 mm, with an increase of no more than 10%, which was considered to be a reasonable control effect.

7.4.2 Cooling Rate Analysis

Comparing the melt pool temperatures and widths at three powers in Fig. 7.7a, it can be found that the instantaneous temperatures of the melt pool after the introduction of ISTC are all significantly reduced, with the largest reduction in temperature at 2400 W. Because of the same control aim of the melt pool width, when the laser power increases, the greater the thermal accumulation suffered by the component. The thermal accumulation effect need to be mitigated by reducing more thermal input (laser power) to further avoid the continuous growth of melt pool width during the ISTC process. The results showed that the ISTC process had significant advantages for solving the problem of melt pool width consistency, especially at high power. As shown in Fig. 7.7b, when deposited the fifth layer in the conventional mode, the melt pool width has already shown an increasing trend compared to the first layer, and it kept increasing. The increase of the melt pool width is also proportional to the laser power [30]. After introduced the ISTC process, the melt pool width remained stable and did not increase by more than 10% at the end. This indicates that the ISTC process is useful in ensuring the stability of the melt pool width of the deposited components at different powers.

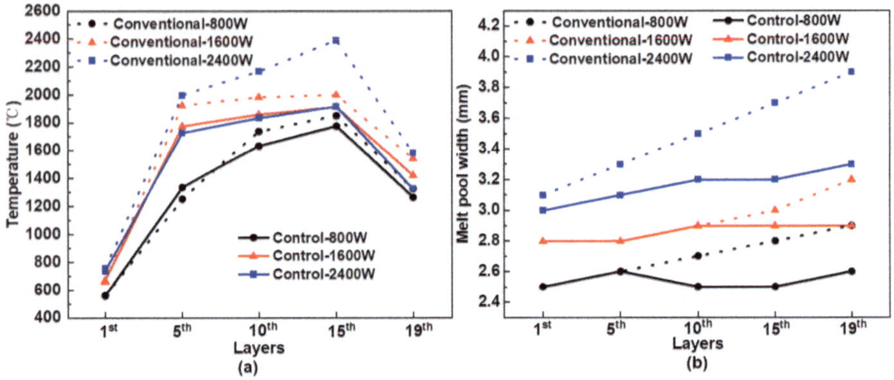

Fig. 7.7 Comparison of melt pool temperature and width at different laser powers. (**a**) temperature, (**b**) width

7.4.3 Surface Roughness Analysis

Figure 7.8 shows the physical drawings and surface morphology of curved-surface thin-walled parts deposited in both modes. When the laser power was 800 W, serious unevenness and spheroidization can be found on the surface of the component, with the surface roughness of ~35.54 μm. In the control mode, the spheroidization still existed, and the surface roughness was only reduced to ~33.26 μm. When the laser power was 1600 W, the deposited components had a relatively clean surface. The surface roughness of the specimens in the conventional mode was ~20.23 μm, and after introducing the ISTC process, the surface roughness decreased to ~18.72 μm. The most significant control effect was achieved at 2400 W. In the conventional mode, the microscope showed unclear surface corrugation due to incomplete solidification of the melt pool, and the surface roughness was ~33.25 μm. In the control mode, the interlayer laps were clear and the corrugation was regular, and the surface roughness was ~21.52 μm, which was a decrease of 35.3% compared with the conventional mode. The results showed that the surface roughness of curved-surface thin-walled parts was most significantly improved by the ISTC process when the laser power was 2400 W. This suggests the high roughness on the surface of the components at high power (2400 W) due to incomplete solidification of the melt pool, which is related to heat accumulation and non-uniformity of the melt pool width. In contrast, at low power (800 W), the spheroidization on the surface of the components was not caused by the thermal accumulation phenomenon.

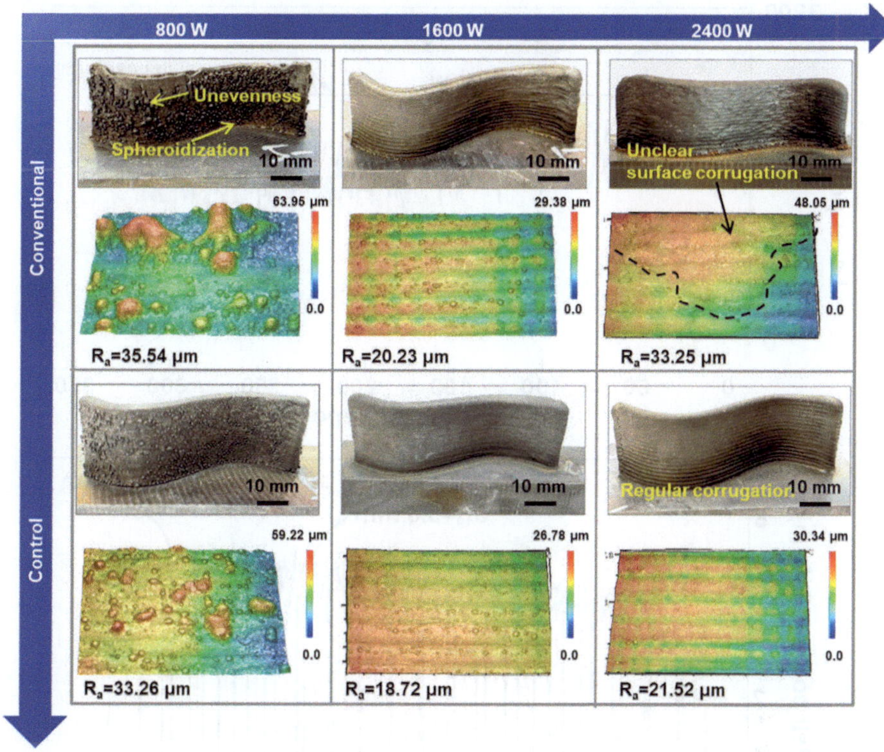

Fig. 7.8 Comparison of surface roughness and morphology of the components at three powers in the conventional mode and control mode

7.5 ISTC of Multiple Curvatures for Manufacturing Curved Components

7.5.1 Melt Pool Width Curves

In Fig. 7.3a, when the laser power was 2400 W, there was a difference in the roughness of the specimen surface due to different curvatures. According to the result in Fig. 7.3c, the surface roughness of the specimen increased significantly when the curvatures were k-K_5. It is presumed that there is a relationship between the heat accumulation and curvatures of the specimens [31]. As shown in Fig. 7.9a, b, the laser powers in the conventional mode were set to the constant 2400 W. The melt pool width increased with the deposition time at different curvatures. At the initial stage of deposition, the melt pool widths were all less than 3 mm. However, at the end of deposition, the melt pool widths were detected to increase to 3.8 mm, 3.9 mm, and 4.2 mm for the K_3, K_4, and K_5 members, respectively. The power curves after ISTC are shown in Fig. 7.9a, to mitigate the thermal accumulation effect, the laser power of different specimens gradually reduced from the initial 2400 W to 1288 W,

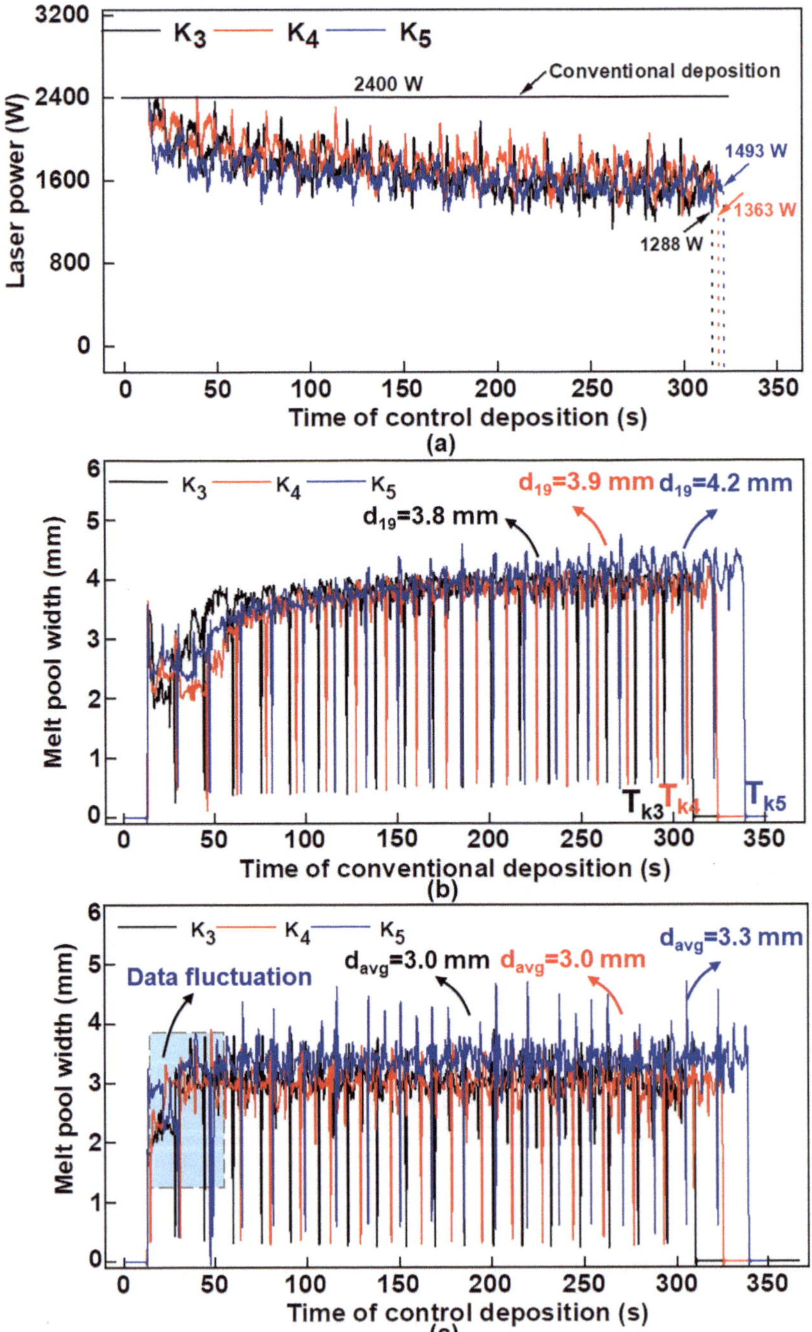

Fig. 7.9 Laser power and melt pool width for two modes with different curvatures. (**a**) Comparison of laser power, (**b**) melt pool width in the conventional mode, (**c**) melt pool width in the control mode

1363 W and 1493 W. In the control mode, as the curvature increased, the melt pool widths at the end of deposition were 3 mm, 3 mm, and 3.3 mm, respectively. When the curvature of the component was K_5, the increase of the melt pool width was less than 10%, which was in line with the expected control target overall.

7.5.2 Cooling Rate Analysis

It can be found that though deposited at the same power, there remains the difference in the melt pool width due to different curvature. In order to investigate the relationship between the curvatures and melt pool width, the temperature gradient and cooling rate at the maximum curvature of the specimens were analysed. The location at the maximum curvature in the scanning path was shown in Fig. 7.10a, and the differences between three curvatures were displayed in Fig. 7.10b. The melt pool temperature gradient at the maximum curvature was shown in Fig. 7.10c. In the conventional mode, the melt pool temperature showed cyclic variations and the solidus and liquidus temperatures of AISI 316L SS were ~1380 °C and ~1430 °C, respectively.

The maximum temperature of the melt pool at the highest curvature was monitored separately during the deposition process. It is notable that the peak temperature was not reached immediately after the material crossed the liquidus point; rather, it resulted from heat accumulation in the preceding layers. As the curvature increased, the maximum temperature of the melt pool also rose, consequently extending the melt pool's lifetime and solidification time. This presented a risk of melt pool collapse and impacted the stability of the manufacturing process [32]. With the implementation of the In-situ Thermal Control (ISTC), the peak temperature of the melt pool was significantly reduced compared to the conventional mode, yet it still demonstrated a positive correlation with increasing curvature. When the curvature increased from K_3 to K_5, the peak temperatures of the melt pool decreased by 2.4%, 13.3% and 19.6%, respectively. The temperature of the melt pool, influenced by the laser power input, decreased in response to the curvature increase, serving as a means to counteract the thermal accumulation effect.

The heating and cooling rates of the specimens at different curvatures were shown in Fig. 7.10d. Due to the complex thermal history, the heating and cooling rates at each point varied with time, and the heating rate was much larger than the cooling rate. The peak cooling rate and peak heat rate (blue dots) decreased with increasing curvature due to thermal accumulation. When the curvature increased to K_5, the cooling rate decreased to ~2808 °C/s and the heating rate increased to −803.5 °C/s. In the ISTC mode, the peak cooling rate was significantly increased compared to that in the conventional mode. Among them, the peak cooling rate and peak heating rate were both the most significant changes at a curvature of K_5, which indicates that the ISTC process is effective in mitigating the thermal accumulation effect of large curvature members.

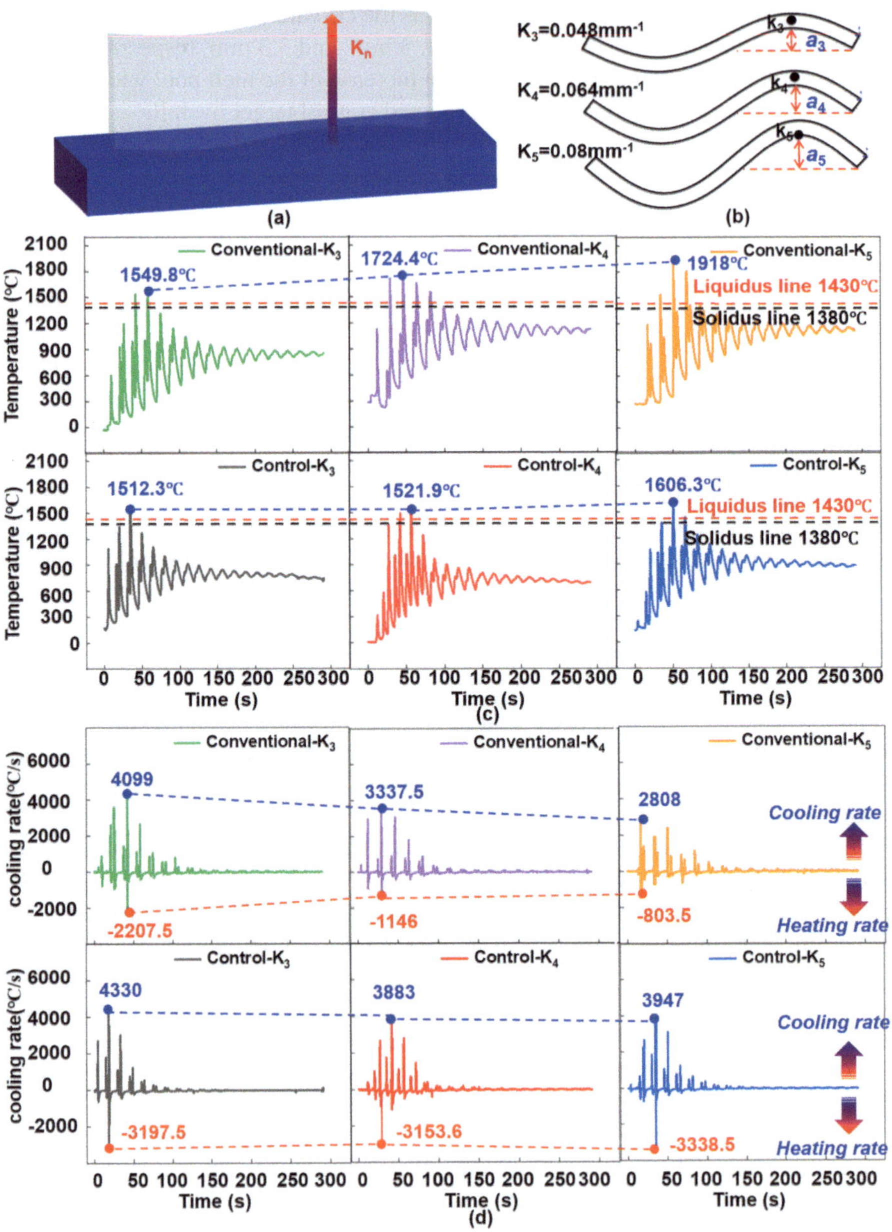

Fig. 7.10 Temperature distribution at the maximum curvature of different parts in the in-situ thermal control mode and conventional mode. (**a**) Location at maximum curvature, (**b**) comparison of the three curvatures, (**c**) thermal circulation temperature, (**d**) cooling rate

7.6 Microstructure and Dimensional Accuracy

7.6.1 Dimensional Accuracy Analysis

As shown in Fig. 7.11, the maximum curvatures of the specimens were chosen to characterise the complete morphology of the cross-section. In the conventional mode (Fig. 7.10a–c), each layer was deposited in two alternating reciprocating tracks, and the overall heights of the curved-surface thin-walled parts deposited with 19 layers were ~23 mm. The initial widths of the melt pool monitored by the camera in Fig. 7.9a were ~3 mm, and the thicknesses of the specimens near the bottom were ~5.8 mm. The thicknesses at the top section of K_3-K_5 were ~6.34 mm, ~6.46 mm, and ~6.62 mm, respectively. As the curvature increased, the thickness of the top of the specimen increased either. In the ISTC mode (Fig. 7.10d–f), the heights were ~26 mm, which substantially improved the uniformity of the overall thickness of the specimen compared to the conventional mode. Three curvatures of the specimens increased from ~5.4 mm, ~5.46 mm, and ~5.5 mm at the bottom to ~5.6 mm, ~5.8 mm, and ~5.9 mm at the top, respectively. The difference in the height of the specimens in two modes is related to the solidification behaviour of the melt pool. By comparing the blue dashed lines of six specimens, it can be observed that the lap rate of two deposition tracks of each layer in the conventional mode was not high. Because of the slow cooling rate, the melt pool is not solidified and cooled in time after its formation. With the increase of the layer numbers, the specimen appeared to collapse, which affected the dimensional accuracy of the overall formation. Under the ISTC process, there were obvious overlap traces between the two melt pools, which indicated that the increases of

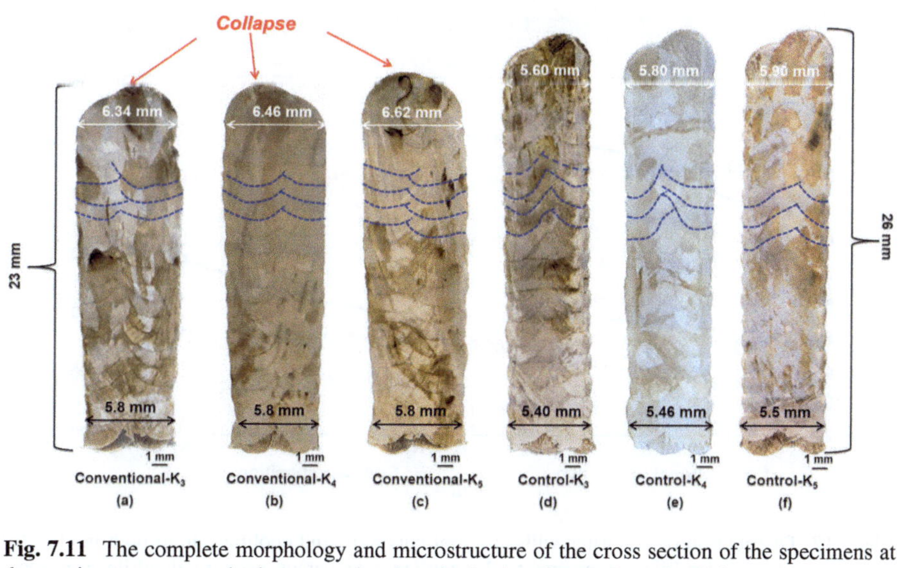

Fig. 7.11 The complete morphology and microstructure of the cross section of the specimens at the maximum curvature in the conventional mode (**a**–**c**) and control mode (**d**–**f**)

solidification rate and cooling rate were conducive to the improvement of the dimensional accuracy of the components.

7.6.2 OM Observation

Further analysis of the microstructures of three components with different curvatures in two deposition modes revealed that they were mainly dominated by columnar dendrites growing along a single direction. The PDAS is an important parameter of the solidification microstructure and has a significant effect on the width of the specimens [33]. As shown in Fig. 7.12a, the PDAS of different specimens were marked with white horizontal lines and dots. The thermal accumulation effect

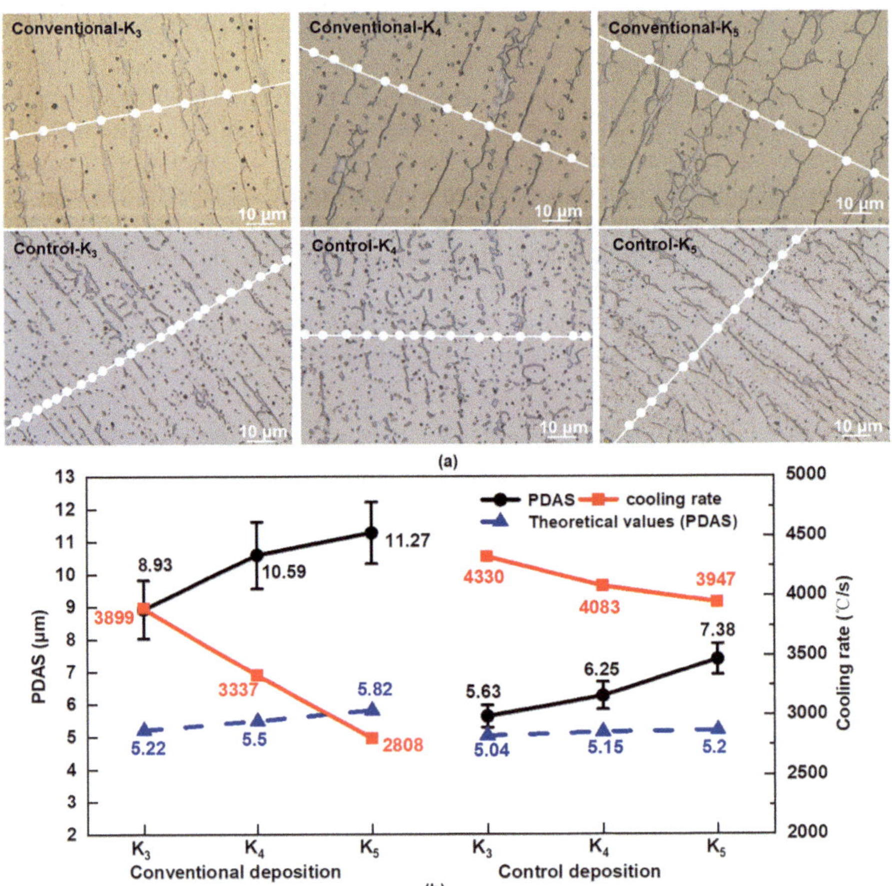

Fig. 7.12 Distribution of dendrites with different curvature and cooling rate in both modes. (**a**) Microstructure of dendrite arrangement, (**b**) cooling rate versus primary dendrite arm spacing (PDAS)

suffered by thin-walled parts deposited at high power is the main factor leading to a decrease in the cooling rate of the melt pool, which further leads to an increase in the PDAS. The statistical data of the PDAS and the cooling rate were exhibited in Fig. 7.12b. In the conventional mode, the cooling rates decreased with increasing curvatures, which were 3899 °C/s, 3337 °C/s, and 2808 °C/s. The PDAS gradually increased from ~8.93 μm to ~11.27 μm. After the ISTC, the cooling rate increased significantly. With the increase of curvature, the overall PDAS showed a gradual increasing trend, but it had been significantly reduced compared with that in the conventional mode, which were ~5.63 μm, ~6.25 μm and ~7.38 μm, respectively. According to the empirical equation between the cooling rate and PDAS of AISI 316L SS [34], it can be expressed as follows:

$$PDAS = \mathrm{a}\dot{T}^{-b} \tag{7.3}$$

where $\dot{T}$ represents the cooling rate, a and b are constants, a ≈ 80 μm (°C/s), and b = 0.33. The theoretical PDAS value calculated according to the empirical equation was shown as the blue dashed line in Fig. 7.12b, and there was a large discrepancy between the actual measured value and the theoretical value (blue line). This indicated that the instantaneous maximum cooling rate in the melt pool cannot be calculated in the theoretical PDAS formula. Because the empirical formula (7.3) mentioned previously is obtained by depositing conventionally shaped (thin walled and block) specimens. Since the shapes of the components in this work were curved, the reason for the discrepancy is that this formula is not applicable to curved thin-walled components. This proves that the three-dimensional geometry of the component is also one of the factors influencing the cooling rate and microstructure distribution. The issue of thermal effects on curved thin-walled parts is covered in the subsequent derivation of the mathematical model [35].

7.6.3 EBSD Observation

As an example of K_5, the EBSD characterizations of the cross-sectional of curved thin-walled parts deposited in the conventional and control modes are shown in Fig. 7.13, where the focus is on the development of the solidification organization in the center of the melt pool, including the inverse pole figure (IPF) map, the phase distribution, and the grain size statistics. As shown in Fig. 7.13a, b, along the accumulation direction of melt pool, it can be seen that there is a clear difference between the growth directions of the large-sized columnar dendrites in two modes in terms of the angle relative to the vertical direction. In the conventional mode, the epitaxial growth angle of the columnar dendrites is ~44°, while in the control mode is ~31°. A larger angle of columnar dendrite epitaxy along the vertical direction implies a larger width of the sample. As can be seen from Fig. 7.13c, d, after removing the zero resolution of the sample preparation, there is no significant difference in the phase occupancy between two modes, which are Fcc-dominated austenitic stainless

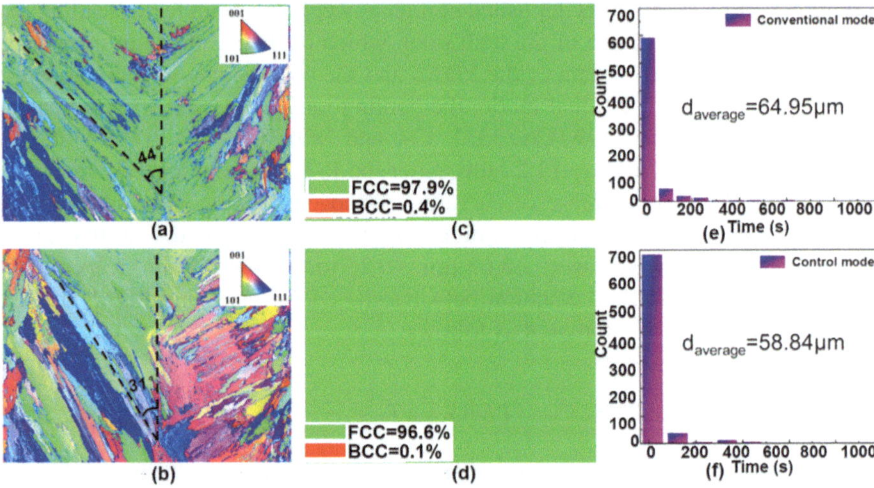

Fig. 7.13 EBSD characterization of deposited curved thin-walled parts in both conventional mode (**a**, **c**, **e**) and control mode (**b**, **d**, **f**). (**a** and **b**) Inverse pole figure orientation maps, (**c** and **d**) phase distribution, (**e** and **f**) grain size statistics

steels. This indicates that there is no significant effect on the phase transition of the 316L stainless steel when only changes the heat input of parts. Based on the statistical grain size results (Fig. 7.13e, f), the average diameter of grains was 58.84 μm in the control mode, which is improved by ~10.4% relative to 64.95 μm in the conventional mode. Based on the EBSD results of the cross-sectional of the curved thin-walled parts, it can be found that the in-situ thermal control technique has a significant effect on the growth direction of the columnar crystals and the grain size.

7.6.4 Effect of ISTC on Forming Accuracy

In this study, the effects of an ISTC process on the deposition of curved-surface thin-walled parts with different powers as well as different curvatures are systematically investigated. Taking the surface roughness and morphology after deposition as an example, the surface roughness is small when the powers are 1200–2000 W in the conventional mode. When the laser powers are 800 W and 2400 W, the surface roughness of the parts are larger. It can be found that the problem of excessive surface roughness is faced at either low or high power. However, it can be seen that the surface morphology of components is various for different powers. At 800 W, the surface of the component shows powder bonding and spheroidisation, which is mostly due to that the power is too low to melt the powder. When the laser power is 2400 W, the surface of the component has a high degree of corrugation and there is no obvious delamination pattern. Because of the high power, the melt pool cannot be cooled in time to solidify and produce. Comparing the ISTC of the initial power of 800 W, 1600 W and 2400 W, it reveals that the best surface roughness enhancement is achieved at 2400 W. According to Fig. 7.3c, when the heat accumulation effect is small (1200–2000 W), the changes in curvatures do not affect the surface roughness. However, when the power increases to 2400 W, the thermal accumulation effect further affects the surface roughness as the curvature of the scanning path increases (K_3–K_5).

The main problems in the LDED of high-power (2400 W) and large-curvature (K_5) thin-walled parts are as follows: low forming accuracy and high surface roughness. According to the heat model established for curved-surface thin-walled parts, the thermal accumulation effect moving with the heat source suffered by the parts is related to the input energy as well as the frequency. Due to the structure and deposition path of parts, the increase in the curvature of the specimen also increases the deposition time in each layer, which leads to an increase in the frequency of the input energy. The high input frequency and high energy results in too low cooling and solidification rates, causing the melt pool to collapse before it has time to cool and solidify. This is related to the cooling rate of the melt pool and the growth of columnar crystals. The cooling rate affects not only the epitaxial growth angle of the columnar crystals but also the PDAS [36]. In general, the typical microstructures of AISI 316L SS are columnar dendrites and equiaxed crystals growing along the normal direction of the melt pool, as well as typical fish scale cell crystals (Fig. 7.14a). However, in the case of curved-surface thin-walled parts, due to the slow cooling rate, the microstructure does not reach the solidification temperature before the next layer is deposited, and the dendrites grow along the direction of heat dissipation from the melt pool and through the melt pool boundary. As shown in Fig. 7.12a, a large number of columnar crystals grow in a single direction. Combined with the EBSD results (Fig. 7.13), it can be found that the epitaxial growth angle of the columnar dendritic crystals is significantly larger in the conventional mode than in the control mode. Because a sharp decrease in the cooling rate encourages excessive epitaxial growth of interlayer dendrites, which limits the growth of equiaxial crystals. According to the relationship between PDAS and cooling rate, a decrease in cooling rate leads to an increase in PDAS, so that a gradual increase in thickness is observed at the cross-section at the maximum curvature of K_5. The epitaxial growth angle of the columnar dendrites and the PDAS together affect the uniformity of the thickness of the component, further reducing the forming accuracy of the part. The temperature coefficient of surface tension in the melt pool affects the surface roughness of the specimen, which is related to the predominant form of convection in the melt pool. During the deposition process, the Marongoni flow gradually flows towards the edge of the melt pool with the accumulation of heat, and the low curvature angle of the melt pool increases with the accumulation of heat [37]. The liquid metal forms a convex shape under tension to reach the equilibrium state, which leads to the uneven surface of the thin-walled part, thus increasing the surface roughness.

According to the thermal accumulation model, the input frequency of curved-surface thin-walled parts is fixed (Fig. 7.14b), so ameliorating the thermal accumulation effect by reducing the laser power is the most direct and effective way by

ISTC process. The input of laser power directly determines the size of the melt pool, and in this study, the melt pool width is used as the control target to reverse the laser power to correct the melt pool width. In order to improve the control accuracy, an adaptive integral separation PID algorithm is used to further reduce the effect of integral overshoot. The improvement of the cooling rate by the ISTC process is significant, and the cooling rate of the K_5 member is increased from 2808 to 3947 °C/s. The extracted melt pool width at the top of the specimen reduced from 3.9 to 3.3 mm. The PDAS, affected by cooling rate, reduced from 5.82 μm to 5.2 μm, which significantly improved the overall forming accuracy of the member (Fig. 7.14c). In the control mode, the heat build-up effect of the melt pool is effectively suppressed. The in-situ thermal control accelerates the cooling rate of the interlayer melt pool, resulting in a large surface temperature gradient. Due to the large temperature gradient and the increase in the lateral thermal gradient, the Marangoni flow activates in the control mode. It flows downwards at the edge of the melt pool wall and upwards in the center of the melt pool in a direction, which causes an increase in the centrifugal velocity of the Marangoni at the sides. The Bond number represents gravity versus surface tension, $Bo = \frac{\rho g h}{2\sigma / R}$, where ρ is the density, g is the gravity, h is the melt pool height, σ is the surface tension, and R is the melt pool radius. For similar melt pool heights (the lift-height set is the same for

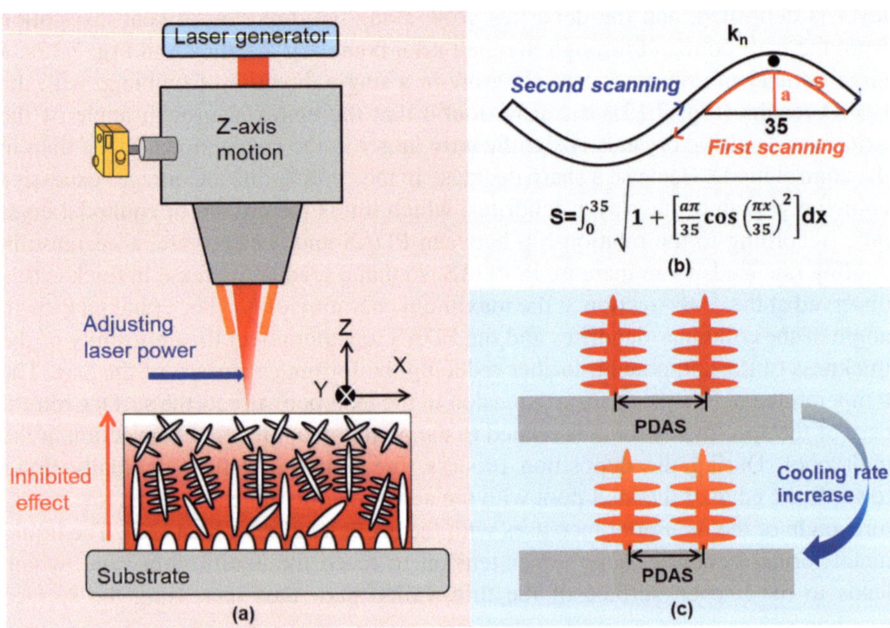

Fig. 7.14 Mechanism of the effects of ISTC techniques on the microstructure. (**a**) The schematic of in-situ thermal control-assisted laser directed energy method, (**b**) scanning path planning, (**c**) effect of cooling rate on primary dendrite arm spacing (PDAS)

both modes), the Bond number decreases as the melt pool radius decreases, suggesting that surface tension is the main factor causing the depression inside the melt pool relative to gravity. The Marangoni flow produces the smoothing effect, which also lead to the reduction of the surface concave-convex morphology and the reduction of the surface roughness.

7.7 Summary

In this chapter, AISI 316L SS curved-surface thin-walled parts with different curvatures and different powers were prepared using LDED assisted by an in-situ thermal control (ISTC) process. A mathematical model of the thermal control effect on the melt pool width was derived, and the effects on dimensional accuracy, microstructure, and tensile properties of curved-surface thin-walled parts were systematically analyzed. The main conclusions were as follows:

1. An original ISTC process based on adaptive integral separation PID algorithm was proposed, which can effectively reduce the surface roughness of the specimen, with the most decrease of 35.3% at 2400 W.
2. The ISTC process mitigated the thermal accumulation effect by increasing the cooling rate and improved the consistency of the melt pool width.
3. The PDAS was further reduced by the cooling rate, promoting a maximum 60.2% improvement in the dimensional accuracy of the components.

References

1. W. Fan, Y.J. Peng, Y. Qi, H. Tan, Z. Feng, Y.X. Wang, F.Y. Zhang, X. Lin, Partially melted powder in laser based directed energy deposition: Formation mechanism and its influence on microstructure, Int. J. Mach. Tool Manufact. 192 (2023) 104072, https://doi.org/10.1016/j.ijmachtools.2023.104072.
2. Z. Sun, W. Guo, L. Li, In-process measurement of melt pool cross-sectional geometry and grain orientation in a laser directed energy deposition additive manufacturing process, Opt. Laser Technol. 129 (2020) 106280, https://doi.org/10.1016/j.optlastec.2020.106280.
3. Y.Y. Su, Z.F. Wang, X. Xu, K.Y. Luo, J.Z. Lu, Effect of closed-loop controlled melt pool width on microstructure and tensile property for Fe-Ni-Cr alloy in directed energy deposition, J. Manuf. Process. 82 (2022) 708–721, https://doi.org/10.1016/j.jmapro.2022.08.049.
4. M. Gharbi, P. Peyre, C. Gorny, M. Carin, S. Morville, P. Le Masson, D. Carron, R. Fabbro, Influence of various process conditions on surface finishes induced by the direct metal deposition laser technique on a Ti–6Al–4V alloy, J. Mater. Process. Technol. 213 (2013) 791–800, https://doi.org/10.1016/j.jmatprotec.2012.11.015.
5. A. Yadollahi, N. Shamsaei, S. Thompson, D. Seely, Effects of process time interval and heat treatment on the mechanical and microstructural properties of direct laser deposited 316L stainless steel, Mater. Sci. Eng. A 644 (2015) 171–183, https://doi.org/10.1016/j.msea.2015.07.056.
6. L. Sinclair, S.J. Clark , Y.H. Chen, S. Marussi, S. Shah, O.V. Magdysyuk, R.C. Atwood, G.J. Baxter, M. Jones, D.G. McCartney, C.L. Alex Leung, P.D. Lee, Sinter formation during

directed energy deposition of titanium alloy powders, Int. J. Mach. Tool Manufact. 176 (2022) 103887, https://doi.org/10.1016/j.ijmachtools.2022.103887.

7. D.K. Kim, W. Woo, J.H. Hwang, K. An, S.H. Choi, Stress partitioning behavior of an AlSi10Mg alloy produced by selective laser melting during tensile deformation using in situ neutron diffraction, J. Alloys Compd. 686 (2016) 281–286, https://doi.org/10.1016/j.jallcom.2016.06.011.
8. K. Kempen, L. Thijs, J. Van Humbeeck, J.P. Kruth, Mechanical properties of AlSi10Mg produced by selective laser melting, Phys. Procedia 39 (2012) 439–446, https://doi.org/10.1016/j.phpro.2012.10.059.
9. Z.C. Yang, X.S. Chen, L.D. Zhu, J.S. Ning, Z.Z. Jiang, M. Yu, S.H. Wang, P.S. Xue, P.H. Xu, B. Xin, Comparative study on forming approaches in directed energy deposition of inclined thin-walled structures, Thin-Walled Struct. 191 (2023) 111074, https://doi.org/10.1016/j.tws.2023.111074.
10. D. W. Patrault, Fast simulation of temperature and phase transitions in directed energy deposition additive manufacturing, Addit. Manuf. 31 (2020) 100990, https://doi.org/10.1016/j.addma.2019.100990.
11. A. C. F. Silveira, R. Fechte-Heinen, J. Epp, Microstructure evolution during laser-directed energy deposition of tool steel by in situ synchrotron X-ray diffraction, Addit. Manuf. 63 (2023) 103408, https://doi.org/10.1016/j.addma.2023.103408.
12. Montevecchi F, Venturini G, Grossi N, Scippa A, Campatelli G. Idle time selection for wire-arc additive manufacturing: A finite element-based technique. Addit Manuf 2018;21:479–86. https://doi.org/10.1016/j.addma.2018.01.007.
13. Ding D, Pan Z, Cuiuri D, Li H. Wire-feed additive manufacturing of metal components: technologies, developments and future interests. Int J Adv Manuf Technol 2015;81:465–81. https://doi.org/10.1007/s00170-015-7077-3
14. A. Heralica, A. Christianssona, B. Lennartson, Height control of laser metal-wire deposition based on iterative learning control and 3D scanning, Opt. Lasers Eng. 50 (9) (2012) 1230–1241, https://doi.org/10.1016/j.optlaseng.2012.03.016.
15. A. Heralica, A. Christianssona, B. Lennartson, Increased stability in laser metal wire deposition through feedback from optical measurements, Opt. Lasers Eng. 48 (2010) 478-485, https://doi.org/10.1016/j.optlaseng.2009.08.012.
16. S. Hwang, W.J. Oh, D.H. Kim, J. G. Kim, J. S. Oh, T.H. Nam, C.S. Kim, T. Lee, Optimizing interlayer cooling for SUS316L thin wall fabricated by directed energy deposition, J. Mater. Res. Technol. 2023; 23: 5239-5245, https://doi.org/10.1016/j.jmrt.2023.02.145.
17. DebRoy T, Wei HL, Zuback JS, Mukherjee T, Elmer JW, Milewski JO, et al. Additive manufacturing of metallic components – Process, structure and properties. Prog Mater Sci 2018;92:112–224. https://doi.org/10.1016/j.pmatsci.2017.10.001.
18. Duan X, Li Q, Xie W, Yang X. Wire arc metal additive manufacturing using pulsed arc plasma (PAP-WAAM) for effective heat management. J Mater Process Technol 2023;311. https://doi.org/10.1016/j.jmatprotec.2022.117806.
19. M. Akbari, R. Kovacevic, An investigation on mechanical and microstructural properties of 316LSi parts fabricated by a robotized laser/wire direct metal deposition system, Addit. Manuf. 23 (2018) 487- 497, https://doi.org/10.1016/j.addma.2018.08.031.
20. Y.Y. Su, Z.F. Wang, H.F. Lu, K.Y. Luo, J.Z. Lu, Improved wear resistance of directed energy deposited Fe-Ni-Cr alloy via closed-loop controlling laser power, J. Manuf. Process. 75 (2022) 802-813, https://doi.org/10.1016/j.jmapro.2022.01.047.
21. I. Jeon, L. Yang, K. Ryu, H. Sohn, Online melt pool depth estimation during directed energy deposition using coaxial infrared camera, laser line scanner, and artificial neural network, Addit. Manuf. 47 (2021) 102295, https://doi.org/10.1016/j.addma.2021.102295.
22. Y. Zheng, C.Z. Li, J.R. Xie, Z.Y. Yu, Y.J. Wang, J.S. Chen, H. Lu, C. Yu, Effects of thermal history of in-situ thermal management on as-built property heterogeneity of plasma arc additively manufactured Inconel 625, J. Mater. Res. Technol. 25 (2023) 2654-2675, https://doi.org/10.1016/j.jmrt.2023.06.090.

23. X. Lin, Q.S. Wang, J.Y.H. Fuh, K.P. Zhu, Motion feature based melt pool monitoring for selective laser melting process, J. Mater. Process. Technol. 303 (2022) 117523, https://doi.org/10.1016/j.jmatprotec.2022.117523.
24. M. H. Farshidianfar, A. Khajepour, A.P. Gerlich, Effect of real-time cooling rate on microstructure in Laser Additive Manufacturing, J. Mater. Process. Technol. 231 (2016) 468-478, https://doi.org/10.1016/j.jmatprotec.2016.01.017.
25. Y.Y Su, C.Y Wang, X. Xu, K.Y. Luo, J.Z. Lu, Pore defects and corrosion behaviour of AISI 316L stainless steel fabricated by laser directed energy deposition under closed-loop control, Surf. Coat. Technol. 463 (2023) 129527, https://doi.org/10.1016/j.surfcoat.2023.129527.
26. Q. Zhong, X.Y. Tian, X.K. Huang , C.B. Huo , D.C. Li, Using feedback control of thermal history to improve quality consistency of parts fabricated via large-scale powder bed fusion, Addit. Manuf. 42 (2021) 101986, https://doi.org/10.1016/j.addma.2021.101986.
27. Z.X. Li, B.H. Chang, K.M. Wang, H.Y. Zhang, Z.Y. Liang, L. Wang, C.M. Liu, D. Du, Closed-loop control of alternating dual-electron beams for molten pool regulation during in-situ additive manufacturing, J. Mater. Process. Technol. 319 (2023) 118087, https://doi.org/10.1016/j.jmatprotec.2023.118087.
28. P. Nandwana, A.M. Elliott, D. Siddel, A. Merriman, W.H. Peter, S.S. Babu, Powder bed binder jet 3D printing of Inconel 718: Densification, microstructural evolution and challenges, Curr. Opin. Solid State Mater. Sci. 21 (2017) 207-218, https://doi.org/10.1016/j.cossms.2016.12.002.
29. S.H. Li, Y.K. Zhao, P. Kumar, U. Ramamurty, Effect of initial dislocation density on the plastic deformation response of 316L stainless steel manufactured by directed energy deposition, Mater. Sci. Eng. A 851 (2022) 143591, https://doi.org/10.1016/j.msea.2022.143591.
30. Y.H. Wang, K.X. Hu, W.D. Li, L.H. Wang, Prediction of melt pool width and layer height for Laser Directed Energy Deposition enabled by physics-driven temporal convolutional network, J. Manuf. Syst. 69 (2023) 1-17, https://doi.org/10.1016/j.jmsy.2023.06.002.
31. P.A. Hooper, Melt pool temperature and cooling rates in laser powder bed fusion, Addit. Manuf. 22 (2018) 548-559, https://doi.org/10.1016/j.addma.2018.05.032.
32. Y. Kong, H.H. Huang, Effect of intrinsic heat treatment on microstructure and hardness of additively manufactured Inconel 625 alloy by directed energy deposition, Mater. Des. 224 (2022) 11279, https://doi.org/10.1016/j.matdes.2022.111279.
33. S. Li, Q. Wei, Y. Shi, Z. Zhu, D. Zhang, Microstructure characteristics of Inconel 625 superalloy manufactured by selective laser melting, J. Mater. Sci. Technol. 31 (9) (2015) 946–952, https://doi.org/10.1016/j.jmst.2014.09.020.
34. B. Zheng, Y. Zhou, J. Smugeresky, J. Schoenung, E. Lavernia, Thermal behavior and microstructure evolution during laser deposition with laser-engineered net shaping: part II. Experimental investigation and discussion, Metall. Mater. Trans. A 39 (2008) 2237–2245, https://doi.org/10.1007/s11661-008-9566-6.
35. Y. Zhou, F. Ning, P. Zhang, A. Sharma, Geometrical, microstructural, and mechanical properties of curved-surface AlSi10Mg parts fabricated by powder bed fusion additive manufacturing, Mater. Des. 198 (2021) 109360, https://doi.org/10.1016/j.matdes.2020.109360.
36. M. Gharbi, P. Peyre, C. Gorny, M. Carin, S. Morville, P. Le Masson, D. Carron, R. Fabbro, Influence of various process conditions on surface finishes induced by the direct metal deposition laser technique on a Ti–6Al–4V alloy, J. Mater. Process. Technol. 213 (2013) 791–800, https://doi.org/10.1016/j.jmatprotec.2012.11.015.
37. Himani, Siva Prasad Frank, Brueckner Alexander F.H., Kaplan (2020) Powder incorporation and spatter formation in high deposition rate blown powder directed energy deposition Additive Manufacturing 35101413, https://doi.org/10.1016/j.addma.2020.101413.

Chapter 8
Thermal Effect Model and Tensile Performance Analysis of Laser Directed Energy Deposition 316L Curved Thin-Walled Parts

8.1 Introduction

Typical aero-engine thin-walled components have complex shapes, and the components are mainly composed of complex curved surfaces. The quality issues of forming thin-walled curved components have attracted widespread attention. Due to the special deposition structure and complex curvatures, as the number of deposition layers increases, the thermal accumulation effect intensifies, and laser deposition thin-walled curved components face serious thermal accumulation problems. Therefore, in Chap. 8, based on Chap. 7, the thermal accumulation effect model of laser directional energy deposition thin-walled curved components was derived, as well as the thermal suppression effect model of thin-walled curved components after introducing in-situ control of the melt pool. In addition, the mechanism of significant enhancement in component toughness due to twin-induced plastic deformation was analyzed, and the promoting effect of thermal regulation on the improvement of component strength and toughness was analyzed.

8.2 Tensile Test Characterization

Tensile experiments were performed on a Zwick Z2.5 TH tensile testing machine equipped with a laserXtens tensiometer. Three parallel specimens per group were used for the experiments.

K. Luo et al., *Melt Pool In-Situ Controlling of Laser Additive Manufacturing Technology*, https://doi.org/10.1007/978-981-95-4597-1_8

8.3 Heat Accumulation Model of the Traditional Mode

The heat source problem for LDED of curved-surface thin-walled parts is the calculation of the temperature distribution of parts under an arbitrarily prescribed laser trajectory, which is the basis for ISTC modelling. In general, the heat source (Q_{heat}) consists mainly of heat absorption (Q_{abs}), heat radiation (Q_{rad}) and heat convection (Q_{conv}) [1, 2]. Compared with the laser heat absorbed by the workpiece, the heat loss due to thermal radiation and thermal convection is small and not directly related to the curvature of the component, so we assume that the workpiece volume is initially in thermal equilibrium with the environment. Considering the workpiece volume as semi-infinite in the z-direction, the heat source model is simplified as

$$Q_{heat} = Q_{abs} = \frac{\eta_{abs} P}{w V_{proc}} \tag{8.1}$$

where η_{abs} is heat absorptivity, P is laser power, w is melt pool width, V_{proc} is velocity.Based on the geometrical properties of the heat source and the resulting dominant heat flow, Carslaw & Jaeger [3] gave explicit solutions to the heat transfer equation. After placing the energy Q_{heat} on a semi-infinite object with density ρ, heat capacity c_p and heat diffusivity, $k = \frac{\lambda}{\rho c_p}$, where λ is the thermal conductivity of the material. The local temperature increases at position r and time t can be described as

$$\Delta T(r,t) = \frac{Q_{heat}}{\rho c_p \sqrt{(4\pi k t)^3}} \cdot e^{-\frac{r^2}{4kt}} \tag{8.2}$$

The heat source (Q_{wall}) of a curved-surfaced thin-walled part is regarded as consisting of multiple layers of accumulated surface heat sources [4]. The heat source

$$Q_{wall} = \frac{2Q_{heat}}{A} \tag{8.3}$$

where the surface heat source is quantified by the ratio of energy to area A, $A = 2lw$, with l being the length of the deposited layer and w being the width of the melt pool. The scanning path in this article can be defined as

$$y = a \sin\left(\frac{\pi x}{35}\right) x \in [0,70] \tag{8.4}$$

where a represents different curvatures, in this experiment a = 2,4,6,8,10. The total length l of the scanning curve is defined as

$$l = 2\int_0^{35} \sqrt{1 + \left[\frac{a\pi}{35}\cos\left(\frac{\pi x}{35}\right)^2\right]} dx \tag{8.5}$$

Thus, the heat source for a curved-surface thin-walled wall is modelled as

$$Q_{wall} = \frac{2\eta_{abs}P}{w^2 l V_{proc}} \tag{8.6}$$

LDED involves multiple successive energy inputs and uses many individual beads or layers to form the component. Each of these individual bead or layer can be considered as a discrete energy input into the component being built. This repeated input of energy occurs with the frequency

$$f = \left(t_{proc} + t_{pause}\right)^{-1} \tag{8.7}$$

where $t_{proc} = N_{bead} \cdot \frac{l_{bead}}{V_{proc}}$ is the time associated with the addition of a layer or bead with the total track length $N_{bead} \cdot l_{bead}$ at the velocity V_{proc}, $N_{bead} = \frac{l}{w}$ is the number of bead, and t_{pause} is the dwell time. In this study, continuous laser energy deposition was used, so the dwell time $t_{pause} = 0$. The Eq. 8.10 further simplifies to

$$f = \frac{wV_{proc}}{l^2} \tag{8.8}$$

Substituting Eq. 8.10 in Eq. 8.9 yields the equation for the heat source with respect to the input frequency (f)

$$Q_{wall} = \frac{2\eta_{abs}P}{wl^3 f} \tag{8.9}$$

When processing materials with ultrashort laser pulses, Weber et al. [5] and Shang et al. [6] proposed to use the local and temporal temperature increase $\Delta T(N_A, r)$ after the sequence of the number of energy inputs N_A,which can be described by the sum $\Delta T_{Residual}(N_A, r)$

$$\Delta T_{Residual}\left(N_A, r\right) = \frac{Q_{wall}}{\rho c_p \sqrt{(4\pi k)^3}} \cdot \sum_{N=1}^{N_A} \frac{1}{\sqrt{(N/f)^3}} \cdot e^{\frac{-f}{N} \cdot \frac{r^2}{4k}} \tag{8.10}$$

The residual temperature difference tends to exist briefly when depositing the next layer, and it is closely related to the value of *f*. That is, as the deposition layer increases, too large the value of f implies the continuous high temperature input, which is prone to trigger the thermal accumulation effect. Substituting Eq. 8.9 into Eq. 8.10 further establishes the thermal accumulation model for curved-surface thin-walled walls:

$$T_{Residual}\left(r, N_A\right) = T_0 + \frac{1}{\rho c_p \sqrt{(4\pi k)^3}} \cdot \sum_{N=1}^{N_A} \frac{2\eta_{abs}P}{wfl^3 \sqrt{(N/f)^3}} \cdot e^{\frac{-f}{N} \cdot \frac{(r-Ns)^2}{4k}} \tag{8.11}$$

In order to further investigate the influence of curvature change on the thermal accumulation effect of the component, the path l of single-pass deposition is decomposed into l_1 and l_2. The scanning path length from the deposition start point to the maximum curvature is set to l_1, and the remaining length is l_2. The heat source equations for different curvature positions for different number of layers are demonstrated as follows

$$\left\{\begin{array}{l} Q_1 = \dfrac{Q_{heat}}{l^1} \\ Q_2 = \dfrac{Q_{heat}}{2l_2} \\ Q_{first} = Q_1 + Q_2 \\ Q_{N_1} = \dfrac{Q_{heat}}{2l^1}(N \geq 2) \\ Q_{N_2} = \dfrac{Q_{heat}}{2l_2} \\ Q_N = Q_{N_1} + Q_{N_2} \end{array}\right\} \tag{8.12}$$

wherein, Q_1 is the heat source at point K at the first maximum curvature of the first pass of the first layer, Q_2 is the heat source at the second pass of the first layer at K, and Q_{first} is the heat source at the end of depositing the first layer at K. Q_{N1} is the heat source at the first pass of the Nth layer at K, Q_{N2} is the heat source at the second pass of the Nth layer at K, and Q_N is the heat source at the end of depositing the Nth layer at K. It is worth noting the range of values of N. Because of the deposition sequence, the heat source models for the first and N_{th} layers are different. The heat source model at the maximum curvature is solved by decomposing the heat source at the maximum curvature

$$\begin{aligned} T_{Residual} &= T_0 + \frac{1}{\rho c_p \sqrt{(4\pi k)^3}} \cdot \left(\frac{Q_{first}}{f^{\left(-\frac{3}{2}\right)}} \cdot e^{-f \cdot \frac{(r-s)^2}{4k}} + \sum_{N=2}^{N_A} \frac{Q_N}{\sqrt{(N/f)^3}} \cdot e^{\frac{-f}{N} \cdot \frac{(r-Ns)^2}{4k}} \right) \\ &= T_0 + \frac{1}{\rho c_p \sqrt{(4\pi k)^3}} \\ &\quad \cdot \left(\frac{2(l_1 + 2l_2)\eta_{abs} P_1}{2l_1 l_2 w l^3 f \sqrt{(1/f)^3}} \cdot e^{-f \cdot \frac{(r-s)^2}{4k}} + \sum_{N=2}^{N_A} \frac{2(l_1 + l_2)\eta_{abs} P_N}{2l_1 l_2 w l^3 f \sqrt{(N/f)^3}} \cdot e^{\frac{-f}{N} \cdot \frac{(r-Ns)^2}{4k}} \right) \end{aligned} \tag{8.13}$$

8.4 Heat Accumulation Model of the Control Mode

The ISTC process takes the melt pool width as the control target, and mitigates the thermal accumulation effect by changing the constant laser power in conventional LDED to the dynamic laser power variation. With the introduction of ISTC, the amount of laser power variation at each time step ΔP is shown. The laser power input is

$$\begin{cases} 1^{bead}: P_1 \sim w_1(k) \\ e_1(k) = |w_s(k) - w_1(k)| \\ \Delta P_1 = K_P \left\{ e_1(k) + \frac{1}{T_I} \sum_{m=1}^{k} e_1(m) + T_D \left[e_1(k) - e_1(k-1) \right] \right\} \\ P_2 = P_1 + \Delta P_1 \\ \ldots\ldots \\ N^{bead}: P_N \sim w_N(k) \\ \Delta P_{\mathrm{N}} = K_P \left\{ e_N(k) + \frac{1}{T_I} \sum_{m=1}^{k} e_N(m) + T_D \left[e_N(k) - e_N(k-1) \right] \right\} \\ P_N = P_{N-1} + \Delta P_{N-1} = P_1 + \sum_{i=1}^{N-1} \Delta P_i \end{cases} \tag{8.14}$$

Wherein, in the control mode deposition stage, 1^{bead} is the number of the first melt pool, P_1 represents the initial power, $w_1(k)$ represents the actual width of the melt pool under the irradiation of 2400 W. $w_s(k)$ represents the width of the reference melt pool, $e_1(k)$ is the difference between the actual width of the first melt pool and the $w_s(k)$, and is the input to the PID controller. ΔP_1 is the amount of change in the laser power corrected according to the width of the first melt pool and the width of the reference value, which is also the output of the PID controller, and P_2 is the input of the second laser power in the control mode (the derivation here including the integral action). When depositing the Nth melt pool, the corresponding parameter settings are similar to those of the first layer. The whole of in-situ melt pool control is a process of continuously correcting the laser power with the reference width of the melt pool as the control target.

Therefore, the thermal accumulation model under ISTC is

$$\begin{aligned} T_{\mathrm{Residual}} = T_0 + \frac{1}{\rho c_p \sqrt{(4\pi k)^3}} \cdot \Bigg(& \frac{2(l_1 + 2l_2)\eta_{abs} P_1}{2l_1 l_2 w l^3 f \sqrt{(1/f)^3}} \cdot e^{-f \cdot \frac{(r-s)^2}{4k}} \\ & + \sum_{N=2}^{N_A} \frac{2(l_1 + l_2)\eta_{abs} P_N}{2l_1 l_2 w l^3 f \sqrt{(N/f)^3}} \cdot e^{\frac{-f}{N} \cdot \frac{(r-Ns)^2}{4k}} \Bigg) \end{aligned} \tag{8.15}$$

where $P_1 = \Delta P + P_0$, P_0 is the original set power of the deposition process. The power at the deposition of the N_{th} layer is

$$P_N = K_P\left\{w(k)+\frac{1}{T_I}\sum_{m=1}^{k}w(m)+T_D\left[w(k)-w(k-1)\right]\right\}+P_{N-1} \tag{8.16}$$

After substituting P_1 and P_N into Eq. 8.17, the heat accumulation model under ISTC is further simplified as

$$\begin{aligned}
T_{\mathrm{Residual}} &= T_0 + \frac{1}{\rho\cdot c_p\cdot\sqrt{(4\pi k)^3}} \\
&\cdot\left(\frac{2(l_1+2l_2)\eta_{abs}(P_0+\Delta P)}{2l_1l_2wl^3f\sqrt{(1/f)^3}}\cdot e^{-f\cdot\frac{(r-s)^2}{4k}}+\sum_{N=2}^{N_A}\frac{2(l_1+l_2)\eta_{abs}(P_{N-1}+\Delta P)}{2l_1l_2wl^3f\sqrt{(N/f)^3}}\cdot e^{\frac{-f}{N}\cdot\frac{(r-Ns)^2}{4k}}\right) \\
&= T_0 + \frac{1}{\rho c_p\sqrt{(4\pi k)^3}} \\
&\cdot\Bigg(\frac{2(l_1+2l_2)\eta_{abs}\left(P_0+K_P\left\{w(k)+\frac{1}{T_I}\sum_{m=1}^{k}w(m)+T_D\left[w(k)-w(k-1)\right]\right\}\right)}{2l_1l_2wl^3f\sqrt{(1/f)^3}} \\
&\cdot e^{-f\cdot\frac{(r-s)^2}{4k}} \\
&+\sum_{N=2}^{N_A}\frac{2(l_1+l_2)\eta_{abs}\left(P_{N-1}+K_P\left\{w(k)+\frac{1}{T_I}\sum_{m=1}^{k}w(m)+T_D\left[w(k)-w(k-1)\right]\right\}\right)}{2l_1l_2wl^3f\sqrt{(N/f)^3}} \\
&\cdot e^{\frac{-f}{N}\cdot\frac{(r-Ns)^2}{4k}}\Bigg)
\end{aligned} \tag{8.17}$$

In order to compare the effect of curvature on the thermal accumulation effect of the parts, the derivation of a model for planar thin-walled parts is also necessary. When the deposition path is a straight line, $l_{straight} = 70$. Equation 8.8 represents a generalized model for depositing the component with a known scanning path length, with the scanning length denoted by l in the equation. Therefore, substituting into Eq. 8.13 yields the heat accumulation model for a planar thin-walled part as

$$T_{\mathrm{Residual}}(r,N_A) = T_0 + \frac{1}{\rho c_p\sqrt{(4\pi k)^3}}\cdot\sum_{N=1}^{N_A}\frac{2\eta_{abs}P}{70^3wf\sqrt{(N/f)^3}}\cdot e^{\frac{-f}{N}\cdot\frac{(r-Ns)^2}{4k}} \tag{8.18}$$

Equation 8.5 shows the curved scanning length in the experiment. Using the curvature K_5 as an example, for Eq. 8.5, the deposition length l_{curve} of the entire member is calculated as

$$l_{\text{curve}} = 2\int_0^{35} \sqrt{1 + \left[\frac{10\pi}{35}\cos\left(\frac{\pi x}{35}\right)\right]^2}\, dx \tag{8.19}$$

substituting into Eq. 8.11 yields the heat accumulation model for a K_5 curved thin-walled part as

$$T_{\text{Residual}}(r, N_A) = T_0 + \frac{1}{\rho c_p \sqrt{(4\pi k)^3}} \cdot \sum_{N=1}^{N_A} \frac{2\eta_{abs} P}{l_{curve}^{\,3} wf \sqrt{(N/f)^3}} \cdot e^{\frac{-f}{N} \cdot \frac{(r-Ns)^2}{4k}} \tag{8.20}$$

Between two points, the straight line path is the shortest. The greater the curvature, the longer the scanning length of the part. Therefore, the thermal accumulation effect is greater for curved thin-walled parts compared to the planar parts.

When deriving the thermal accumulation model, the r coordinate system was introduced to obtain the residual temperature of the melt pool increasing with deposition time and scanning position. Therefore, when depositing other structure (bulk structure, overhanging structure), the thermal accumulation effect model can be similarly obtained by transforming the 3D coordinate system of the components into the r coordinate system. The thermal accumulation effect of curved-surface thin-walled parts provides a reference for other shapes, and the heat effect model of different kinds of parts will be the focus of the future research.

8.5 Tensile Property

8.5.1 Engineering Stress-Strain Curves

The tensile specimen was selected at the maximum curvature of the parts along the deposition direction, and the specimen dimension was also displayed. Figure 8.1a, b respectively showed the stress-strain curves and statistical histograms for three curvatures in two deposition modes. It could be noticed that the strength and plasticity of the specimens prepared by the ISTC process were increased compared with the conventional mode, especially the elongation. Specifically, in Fig. 8.1b, the tensile strengths were increased by 3 MPa, 13 MPa, and 1 MPa. The elongations were enhanced by 14.7%, 9% and 38.75% when the curvatures were K_3–K_5, respectively. It was noteworthy that quasi-superplasticity (elongation over 100%) was achieved in all specimens after ISTC. By comparing the tensile strength and elongation of 316L stainless steel fabricated by additive manufacturing in other publications, it can be seen that the elongation of 316L fabricated by in-situ thermal control-assisted LDED is much higher than that of the conventional fabrication method (Table 8.1).

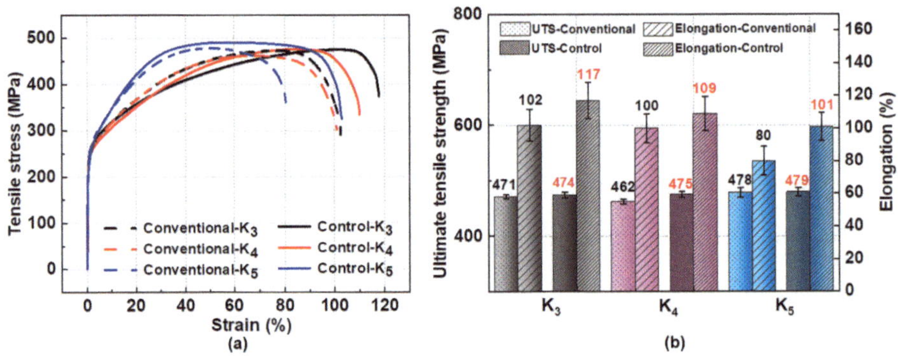

Fig. 8.1 Tensile properties analysis of different curvatures in two deposition modes. (**a**) Tensile curves, (**b**) comparison of tensile strength and elongation

Table 8.1 Comparative data on tensile properties of 316L stainless steel with other publications [7–14]

Deposition method	UTS (MPa)	EL (%)
LDED [7]	600	20
LDED [8]	500	45
LDED [9]	600	40
LDED [10]	700	25
LPBF [11]	650	12
LPBF [12]	598	30
WAAM [13]	500	48
WAAM [14]	450	60
Our work	474	117

UTS ultimate tensile strength, *EL* elongation, *LDED* laser directed energy deposition, *LPBF* laser powder bed fusion, *WAAM* Wire Arc Additive Manufacture

8.5.2 Fractural Morphology

The fractural morphology of the parts with different curvatures was shown in Fig. 8.2. The fracture surfaces of all specimens were not regular, representing that all of them experienced a severe stage of necking deformation during the tensile process, which was verified by the morphology of the fracture edges. In the conventional mode, a number of pores distributed along the edges were found on the fracture surfaces, with the highest ones found on the fracture surfaces when the curvature was K_5. The formation of the pores was attributed to the poor metallurgical bonding between the layers, which seriously affected the elongation. No significant distribution of pores was found on the fracture surfaces of the specimens prepared under the ISTC process. The evolution patterns from the port edges to the centre are necking morphology, shear dimple and equiaxed dimple network, as shown in Fig. 8.2c. The enlarged fracture surface features are displayed in Fig. 8.2d. The diameters of the pores at the fracture surface were all less than 1 μm when the curvature was K_3. The diameter of the pores gradually increased, up to 3 μm at the curvature of K_5. The fracture surface was mainly dominated by the dense dimple network under the ISTC.

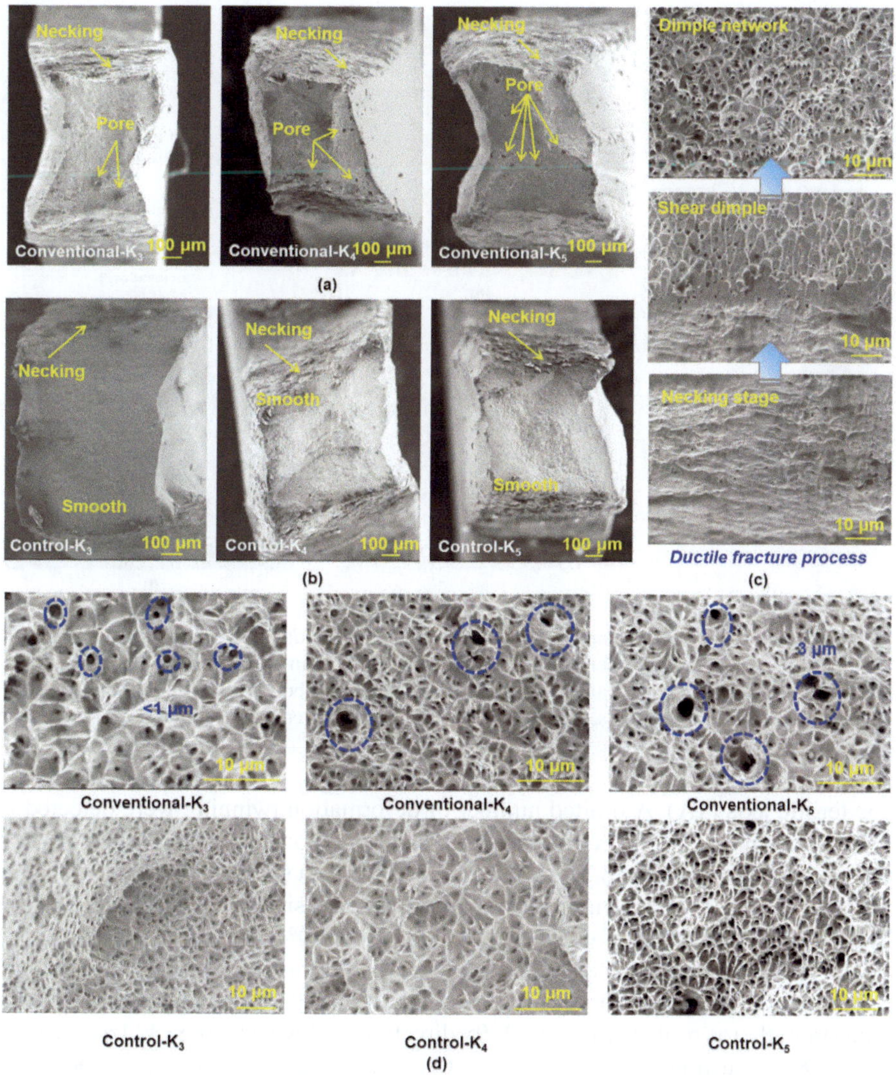

Fig. 8.2 Fractural morphology corresponding to the three curvature components in different deposition modes. (**a**) Conventional mode, (**b**) in-situ thermal control mode, (**c**) process analysis of ductile fracture, (**d**) SEM characterisation of fractural morphology of different curvatures in two deposition modes

8.5.3 EBSD Characterization of Fracture

The tensile direction of the specimen in the conventional mode was shown in Fig. 8.3a, and the position close to the fracture was selected for EBSD analysis. Based on the orientation image in Fig. 8.3b, it can be found that no deformation twinning were formed in the <101> and <001> orientated matrix which parallel to

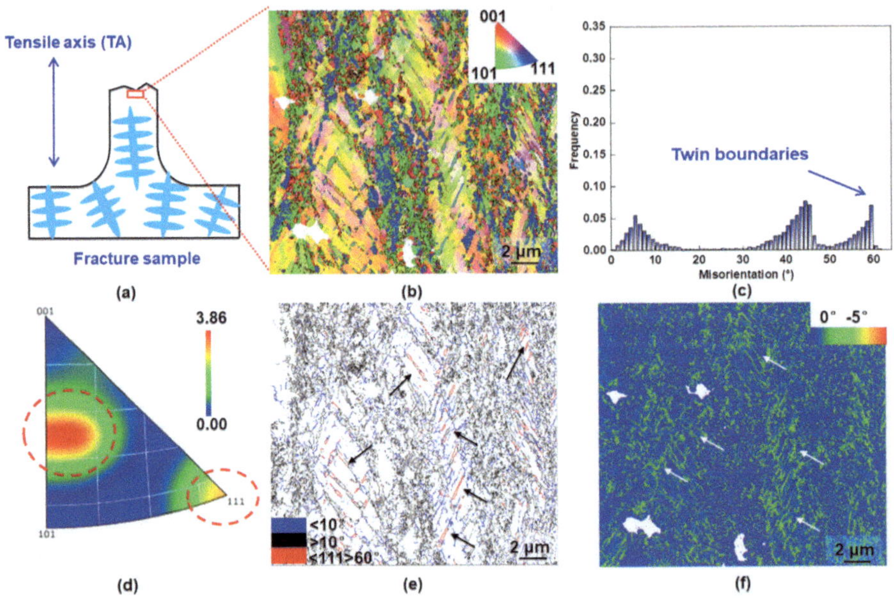

Fig. 8.3 EBSD analysis of samples after fracture in the conventional mode. (**a**) fracture location, (**b**) IPF orientation map, (**c**) orientation angle distribution of grain boundaries, (**d**) the corresponding inverse pole figure (IPF) along the tensile axis, (**e**) distribution of different angular grain boundaries and twin boundaries, (**f**) the corresponding kernel average misorientation (KAM) map

the tensile axis (TA). A limited number of deformation twinning were detected in the grains with <111> || TA orientations. The orientation angle distribution of grain boundaries shown in Fig. 8.3c demonstrated that only a small number of twins were formed in the <111> sample during deformation, since the deformation twin boundaries showed only a very small peak at about 60° and occurred less than 5% of the time. Figure 8.3d showed the inverse pole figure (IPF) in line with TA, presenting the texture after the tensile deformation. In IPF, the highest grain orientation strength was probably in the <210> direction, followed by the <111> direction. Figure 8.3e illustrated the distribution of grain boundaries at different angles, with those less than 10° and greater than 10° marked in blue and black, respectively. In particular, the 60° grain boundaries of the <111> oriented deformation twin were marked in red and pointed out with black arrows. Figure 8.3f depicted the kernel average misorientation (KAM) map, where slip bands were identified as traces with higher local strains (visualised by their higher KAM values, indicated in green). The results indicated that <001> oriented grains were more susceptible to slip deformation.

With ISTC, the tensile direction of the specimen and the sampling location of the EBSD specimen were the same as the conventional direction, shown in Fig. 8.4a. As

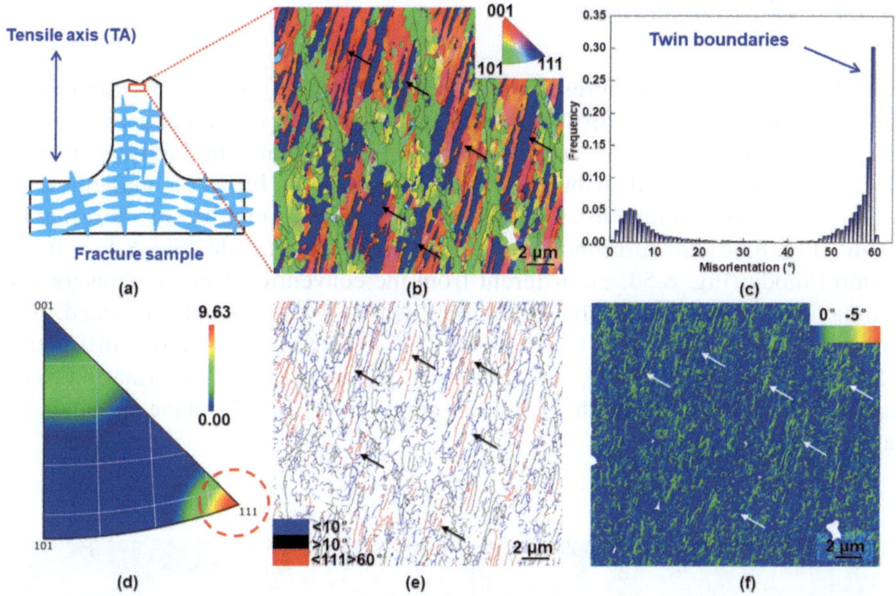

Fig. 8.4 EBSD analysis of samples after fracture in the control mode. (**a**) fracture location, (**b**) IPF orientation map, (**c**) orientation angle distribution of grain boundaries, (**d**) the corresponding inverse pole figure (IPF) along the tensile axis, (**e**) distribution of different angular grain boundaries and twin boundaries, (**f**) the corresponding kernel average misorientation (KAM) map

can be seen in Fig. 8.4b, a large number of <111> deformation twins were generated during tensile deformation, marked by black arrows. The orientation angle distribution of grain boundaries shown in Fig. 8.4c illustrated a peak height at about 60° with a frequency greater than 30%, which proved that a large number of twins were produced during deformation of the <111> sample. In the IPF, the strongest intensity was concentrated in the <111> direction and few grain rotations were observed. The red grain boundaries in Fig. 8.4e represented <111> 60° and were marked with black arrows. The large number of <111> 60° grains represented twin-induced deformation during tensile. This was also verified by the KAM plot in Fig. 8.4f. There is a correlation between deformation twinning activity and KAM [15]. In regions where twinning is more active, the average orientation error is smaller. In particular, strong average misorientation was found in grains (<001> || TA) where there were no thick twins but strong dislocation slip. Twin boundaries acted as barriers to dislocation motion, effectively reducing the mean free path of dislocations and creating a "dynamic Hall-Petch" effect [16–18]. It can increase the strain-hardening capacity of the material during deformation, leading to the twinning induced plasticity (TWIP) effect.

8.5.4 TEM Observation

The TEM characterization results in Fig. 8.5 further demonstrate the significant difference between the deformation twin and dislocation structures in two modes. As shown in Fig. 8.5a–c, the traditional mode is dominated by high-density dislocations and sparsely distributed deformation twins. In Fig. 8.5a, magnification of the structure within the circle reveals that a small number of deformation twins (Fig. 8.5b) are surrounded by a large dislocation tangle (Fig. 8.5c). In the control mode (Fig. 8.5d, e), different from the conventional mode, densely distributed twins are found. The dislocation density is significantly reduced, promoting more deformation twins. From the selected area electron diffraction (SAED) pattern in Fig. 8.5f, a total of two different planes of deformation twins are found, clustered along the γ {0 0 2} plane and γ {0 2–2} plane (γ stands for austenite), respectively.

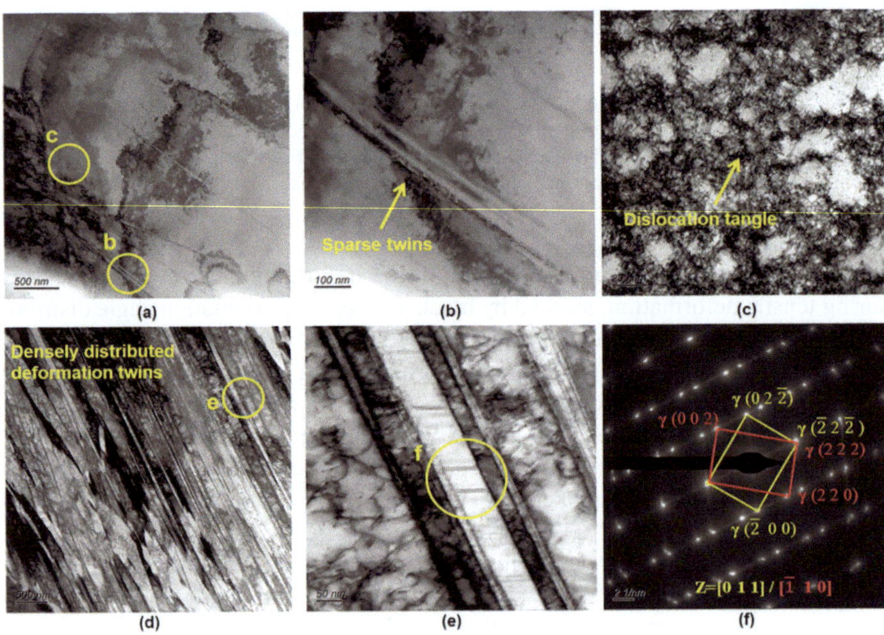

Fig. 8.5 The transmission electron microscope (TEM) characterization of dislocation structure and twinning in both modes. (**a**) The microstructure of the specimen in the conventional mode is dominated by (**b**) sparsely distributed twins and (**c**) high-density dislocation tangles, while (**d**) in the control mode is (**e**) densely distributed deformation twins. (**f**) Selected area electron diffraction (SAED) pattern of deformed twins

8.5.5 Fractural Mechanism

According to the stress-strain curves and statistical plots, the elongation in the conventional mode was greater than or equal to 100% when the curvatures were K_3 and K_4. However, when the curvature increased to K_5, the elongation was only 80%. By analysing the fractural morphology, several densely distributed pores were found in the conventional mode. As the curvature increases to K_5, the size and density of the pores increase, further affecting the interlayer metallurgical bonding of the specimen. When the oxygen content inside the deposition chamber is high, the melt pool will be oxidized and a high melting point oxide film will be formed in the high temperature region which is not protected by the protective gas [19]. When the next layer is deposited on the oxidized surface, the surface tension between the liquid melt pool and the oxide film is low due to the high melting point of the oxide, which promotes the instability of the melt pool [20]. So the liquid metal could not easily spread on its surface, but easy to form a ball and result in a poor formation or spatters. In this experiment, the defect types were mainly small-sized pores, which were quite different from those caused by high oxygen levels in the chamber. Since only the input laser power was changed in both deposition modes, but not the gas environment in the chamber, which suggests that the differences in the porosity were mainly caused by the thermal accumulation effect on the temperature gradient and mobility of the melt pool. According to Tan et al. [21], the melt pool temperature affected the escape velocity of the gas in the melt pool. When the melt pool temperature increases, the rapidly vaporizing water vapor compresses the liquid surface, and the gas bubbles have no time to escape and are trapped in the sedimentary layer, resulting in the formation of pores. In addition, Zheng et al. [22] demonstrated that, under the influence of the surface tension and temperature gradient of the melt pool, the shear force generated during the melt pool flow reduced the solidification rate of the liquid metal and led to the collapse of the liquid metal, resulting in the formation of pores. Under the ISTC process, the plasticity of the components was all effectively improved. The cooling rate enhancement of the members was most significant when the curvature was K_5, and therefore the elongation was also increased the most, from 80% to 101%. By increasing the cooling rate, it is possible to reduce the heat accumulation so that more water vapour is not vaporised, thus reducing pore defects. As can be seen from Fig. 8.2b, there are no obvious pores on the fracture surface in the control mode.

Due to the small size and relatively sparse distribution of pores, the inhibition of pores by ISTC is not the main reason for the plasticity improvement [23]. By comparing the EBSD results in Figs. 8.3 and 8.4, it can be found that more deformation twins produced in the components under ISTC compared to the control mode. The generation of deformation twins is related to the stacking fault energy (SFE) [24]. Figure 8.6 presents the mechanism of ISTC-induced twinning plastic deformation of curved-surface thin-walled parts. The low SFE inhibits cross-slip and instead promotes planar slip and twinning. The suppressed cross-slip leads to an increase in dislocation storage, which together with the emerging twin boundaries increases the

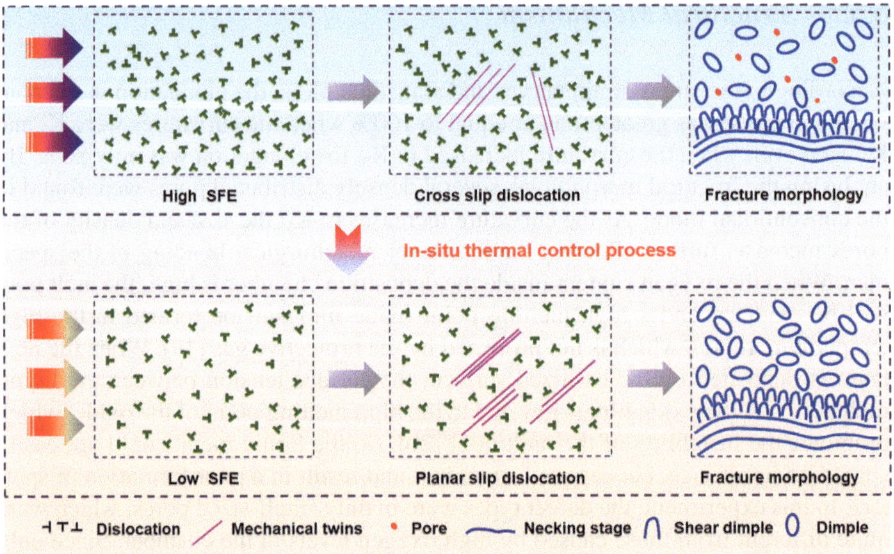

Fig. 8.6 Mechanism of in-situ thermal control induced twinning plastic deformation of curved-surface thin-walled parts

strain hardening rate. Dislocations cause misalignment of atoms within the crystal, which in turn causes elastic distortion of the crystal point structure in its vicinity. SFE refers to the fact that when a complete dislocation breaks down on a crystal face, the normal point structure between the parts of the dislocation on that face is further disrupted, contributing to an increase in the energy of the system [25]. SFE is closely related to the dislocation density. The dislocations of materials will be more likely form at high temperatures, increasing the dislocation density [26]. The ISTC process further promotes more deformation twins by reducing the dislocation density and SFE through decreasing the thermal accumulation effect. Therefore, the twinning-induced plastic (TWIP) deformation effect is the main reason for the quasi-superplasticity of AISI 316L SS deposited with LDED assisted by ISTC.

In addition, it can be found that the ISTC process does not enhance the tensile strength of the material significantly. In general, for AISI 316L SS, there are usually two ways to increase the tensile strength: increasing the dislocation density and decreasing the grain size [27, 28] In the current study, it can be found that the introduction of ISTC reduces the dislocation density and promotes more deformation twins, which hinders dislocation slip. Moreover, due to the specificity of the deposition path of curved-surface thin-walled parts, the overall grains are mostly columnar dendrites growing across the melt pool boundary with large grain sizes. The ISTC process can increase the cooling rate and solidification rate of the melt pool by mitigating the heat accumulation effect, thus reducing the grain size to a certain extent [29]. In this case, the dislocation density and grain size decrease simultaneously in the control mode, so the overall change in the tensile strength of the material is not obvious. The ISTC process demonstrates the possibility of significantly enhancing the plasticity of the material while preserving its tensile strength, contributing to the improvement of the overall material properties.

8.6 Summary

In this chapter, based on the content of Chap. 7, the influence of the combined control of melt pool width and temperature on the tensile properties of laser-deposited 316L stainless steel curved thin-walled parts was further analyzed. The heat accumulation model under the traditional mode and the heat suppression model under the control mode were respectively derived. The plastic deformation effect induced by twinning under thermal control was discovered. The tensile components exhibited quasi-ultrapolymorphism. The main conclusions are as follows:

1. Based on the classical heat source theory model, thermal effect models for the curvature of components and laser power were derived under two modes.
2. The ISTC process further induced twinning plastic deformation by reducing the thermal effect and dislocation density, and the elongation was increased by up to 38.75%. Quasi-superplasticity was achieved in all specimens after control.
3. The ISTC process has contributed to solving the thermal accumulation effect problems, and improving the forming quality and overall performance of curved-surface thin-walled parts. The corresponding mathematical model developed provides a new reference for future additive manufacturing of other metallic materials and different sizes components in this field.

References

1. N. Rykalin, Calculations of Thermal Processes in Welding, Mashgiz, Moscow, 1960.
2. H. Peng, M. Ghasri-Khouzani, S. Gong, R. Attardo, P. Ostiguy, B.A. Gatrell, et al., Fast prediction of thermal distortion in metal powder bed fusion additive manufacturing: Part 1, a thermal circuit network model, Addit. Manuf. 22 (2018), 852–68, https://doi.org/10.1016/j.addma.2018.05.023.
3. H.S. Carslaw, J.C. Jaeger. Conduction of Heat in Solids 1947, second ed., Oxford University Press, Oxford, 1959.
4. C. Hagenlocher, P. O'Toole, W. Xu, M. Brandt, M. Easton, A. Molotnikov, Analytical modelling of heat accumulation in laser based additive manufacturing processes of metals, Addit. Manuf. 60 (2022) 103263, https://doi.org/10.1016/j.addma.2022.103263.
5. R. Weber, T. Graf, P. Berger, V. Onuseit, M. Wiedenmann, C. Freitag, A. Feuer, Heat accumulation during pulsed laser materials processing, Opt. Express 22 (9) (2014) 11312–11324, https://doi.org/10.1364/OE.22.011312.
6. Z.D. Shang, Z.R. Liao, J.A. Sarasua, J. Billinghamd, D. Axinte, On modelling of laser assisted machining: Forward and inverse problems for heat placement control, Int. J. Mach. Tool Manufact. 138 (2019) 36-50, https://doi.org/10.1016/j.ijmachtools.2018.12.001.
7. S. B. Han, Y. S. Lee, S. H. Park, H.J. Song, Ti-containing 316L stainless steels with excellent tensile properties fabricated by directed energy deposition additive manufacturing, Mater. Sci. Eng. A. 862 (2023) 144414, https://doi.org/10.1016/j.msea.2022.144414.

8. S.H. Li, Y.K. Zhao, P. Kumar, U. Ramamurty, Effect of initial dislocation density on the plastic deformation response of 316L stainless steel manufactured by directed energy deposition, Mater. Sci. Eng. A. 851 (2022)143591, https://doi.org/10.1016/j.msea.2022.143591.
9. J.T. Pacheco, V.H. Meura, P. R.A. Bloemer, M.T. Veiga, O.C. M. Filho, A. Cunha, M.F. Teixeira, Laser directed energy deposition of AISI 316L stainless steel: The effect of build direction on mechanical properties in as-built and heat-treated conditions, Adv. Indust. Manuf. Eng. 4 (2022) 100079, https://doi.org/10.1016/j.aime.2022.100079.
10. L. Zhang, W.G. Zhai, W. Zhou, X.Q. Chen, L.J. Chen, B. Han, L.C. Cao, G.J. Bi, Improvement of mechanical properties through inhibition of oxidation by adding TiC particles in laser aided additive manufacturing of stainless steel 316L, Mater. Sci. Eng. A. 853 (2022) 143767, https://doi.org/10.1016/j.msea.2022.143767.
11. W.W. Deng, H.F. Lu, Y.H. Xing, K.Y. Luo, J.Z. Lu, Effect of laser shock peening on tensile properties and microstructure of selective laser melted 316L stainless steel with different build directions, Mater. Sci. Eng. A. 850 (2022) 143567, https://doi.org/10.1016/j.msea.2022.143567.
12. K. Somlo, S.S. Chauhan, C.F. Niordson, K. Poulios, Uniaxial tensile behaviour of additively manufactured elastically isotropic truss lattices made of 316L, Int. J. solid. Struct. 246-247 (2022) 111599, https://doi.org/10.1016/j.ijsolstr.2022.111599.
13. C. Wang, T.G. Liu, P. Zhu, Y.H. Lu, T. Shoji, Study on microstructure and tensile properties of 316L stainless steel fabricated by CMT wire and arc additive manufacturing, Mater. Sci. Eng. A. 769 (2020) 140006, https://doi.org/10.1016/j.msea.2020.140006.
14. C. Wang, P. Zhu, Y.H. Lu, T. Shoji, Effect of heat treatment temperature on microstructure and tensile properties of austenitic stainless 316L using wire and arc additive manufacturing, Mater. Sci. Eng. A. 832 (2022) 142446, https://doi.org/10.1016/j.msea.2021.142446.
15. M.S. Pham, B. Dovgyy, P.A. Hooper, Twinning induced plasticity in austenitic stainless steel 316L made by additive manufacturing, Mater. Sci. Eng. A 704 (2017) 102–111, https://doi.org/10.1016/j.msea.2017.07.082.
16. W.W. Deng, C.Y. Wang, H.F. Lu, X.K. Meng, Z. Wang , J.M. Lv ,K.Y. Luo, J.Z. Lu, Progressive developments, challenges and future trends in laser shock peening of metallic materials and alloys: A comprehensive review, Int. J. Mach. Tool Manufact. 191 (2023) 104061, https://doi.org/10.1016/j.ijmachtools.2023.104061.
17. B.C. De Cooman, Y. Estrin, S.K. Kim, Twinning-induced plasticity (TWIP) steels, Acta Mater. 142 (2018) 283–362, https://doi.org/10.1016/j.actamat.2017.06.046.
18. H. Beladi, I.B. Timokhina, Y. Estrin, J. Kim, B.C. De Cooman, S.K. Kim, Orientation dependence of twinning and strain hardening behaviour of a high manganese twinning induced plasticity steel with polycrystalline structure, Acta Mater. 59 (20) (2011) 7787–7799, https://doi.org/10.1016/j.actamat.2011.08.031.
19. H.S. Prasad, F. Brueckner, A.F.H. Kaplan, Powder incorporation and spatter formation in high deposition rate blown powder directed energy deposition, Addit Manuf 35 (2020), 101413.
20. N. Emminghaus, C. Hoff, J. Hermsdorf, S. Kaierle, Residual oxygen content and powder recycling: Effects on surface roughness and porosity of additively manufactured Ti-6Al-4V, Addit Manuf (2021), 102093.
21. Z. E. Tan, J. H. L. Pang, J. Kaminski, H. Pepin, Characterisation of porosity, density, and microstructure of directed energy deposited stainless steel AISI 316L, Addit. Manuf. 25 (2019) 286-296, https://doi.org/10.1016/j.addma.2018.11.014.
22. B. Zheng, J.C. Haley, N. Yang, J. Yee, K.W. Terrassa, Y. Zhou, E.J. Lavernia, J.M. Schoenung, On the evolution of microstructure and defect control in 316L SS components fabricated via directed energy deposition, Mater. Sci. Eng. A. 764 (2019) 138243, https://doi.org/10.1016/j.msea.2019.138243.
23. R.K. Desu, H. Nitin Krishnamurthy, A. Balu, A.K. Gupta, S.K. Singh, Mechanical properties of Austenitic Stainless Steel 304L and 316L at elevated temperatures, J. Mater. Res. Technol. 5 (2016) 13, https://doi.org/10.1016/j.jmrt.2015.04.001.

24. D. Molnár, X. Sun, S. Lua, W. Lia, G. Engberg, L. Vitos, Effect of temperature on the stacking fault energy and deformation behaviour in 316L austenitic stainless steel, Mater. Sci. Eng. A 759 (2019) 490-497, https://doi.org/10.1016/j.msea.2019.05.079.
25. P. Chowdhury, H. Sehitoglu, W. Abuzaid, H. Maier, Mechanical response of low stacking fault energy co-ni alloys-continuum, mesoscopic and atomic level treatments, Int. J. Plast. 71 (2015) 32-61, https://doi.org/10.1016/j.ijplas.2015.04.003.
26. B.K. Zhao, P. Huang, L.B. Zhang, S.Z. Li, Z. Zhang, Q. Yu, Temperature Efect on Stacking Fault Energy and Deformation Mechanisms in Titanium and Titanium-aluminium Alloy, Sci. Rep. (2020) 10:3086, https://doi.org/10.1038/s41598-020-60013-6.
27. Y. Balit, E. Charkaluk, A. Constantinescu, Digital image correlation for microstructural analysis of deformation pattern in additively manufactured 316L thin walls, Addit. Manuf. 31 (2020) 100862, https://doi.org/10.1016/j.addma.2019.100862.
28. B. Barkia , P. Aubry, P. Haghi-Ashtiani, T. Auger, L. Gosmain, F. Schuster, H. Maskrot, On the origin of the high tensile strength and ductility of additively manufactured 316L stainless steel: Multiscale investigation, J. Mater. Sci. Technol. 41 (2020) 209-218, https://doi.org/10.1016/j.jmst.2019.09.017.
29. A. D. Murphy-Leonard, D. C. Pagan, P. G. Callahan, Z. K. Heinkel, C. E. Jasien, D. J. Rowenhorst, Investigation of porosity, texture, and deformation behaviour using high energy X-rays during in-situ tensile loading in additively manufactured 316L stainless steel, Mater. Sci. Eng. A 810 (2021) 141034, https://doi.org/10.1016/j.msea.2021.141034.

MIX
Papier aus verantwortungsvollen Quellen
Paper from responsible sources
FSC® C105338
FSC
www.fsc.org

If you have any concerns about our products,
you can contact us on
ProductSafety@springernature.com

In case Publisher is established outside the EU,
the EU authorized representative is:
Springer Nature Customer Service Center GmbH
Europaplatz 3, 69115 Heidelberg, Germany

Printed by Libri Plureos GmbH
in Hamburg, Germany